A PRIMER OF STATISTICAL MECHANICS

A PRIMER OF STATISTICAL MECHANICS

R B SINGH
Former Head
Department of Physics, UNPG College
Padrauna, DDU Gorakhpur University
Gorakhpur (UP)

NEW AGE
REFERENCE

An Imprint of

NEW AGE INTERNATIONAL (P) LIMITED, PUBLISHERS
LONDON • NEW DELHI • NAIROBI
Bangalore • Chennai • Cochin • Guwahati • Hyderabad • Kolkata • Lucknow • Mumbai
Visit us at **www.newagepublishers.com**

Published by New Age International (P) Ltd., Publishers
First Edition: 2006
Reprint: 2018

GLOBAL OFFICES

- **New Delhi** **NEW AGE INTERNATIONAL (P) LIMITED, PUBLISHERS**
7/30 A, Daryaganj, New Delhi-110002, (INDIA)
Tel.: (011) 23253771, 23253472, **Telefax:** 23267437, 43551305
E-mail: contactus@newagepublishers.com • Visit us at www.newagepublishers.com
- **London** **NEW AGE INTERNATIONAL (UK) LTD.**
27 Old Gloucester Street, London, WC1N 3AX, UK
E-mail: info@newacademicscience.co.uk • Visit us at www.newacademicscience.co.uk
- **Nairobi** **NEW AGE GOLDEN (EAST AFRICA) LTD.**
Ground Floor, Westlands Arcade, Chiromo Road (Next to Naivas Supermarket), Westlands, Nairobi, KENYA
Tel.: 00-254-713848772, 00-254-725700286
E-mail: kenya@newagepublishers.com

BRANCHES

- **Bangalore** 37/10, 8th Cross (Near Hanuman Temple), Azad Nagar, Chamarajpet, Bangalore- 560 018
Tel.: (080) 26756823, **Telefax:** 26756820, **E-mail: bangalore@newagepublishers.com**
- **Chennai** 26, Damodaran Street, T. Nagar, Chennai-600 017, **Tel.:** (044) 24353401, **Telefax:** 24351463
E-mail: chennai@newagepublishers.com
- **Cochin** CC-39/1016, Carrier Station Road, Ernakulam South, Cochin-682 016, **Tel.:** (0484) 2377303, **Telefax:** 4051304
E-mail: cochin@newagepublishers.com
- **Guwahati** Hemsen Complex, Mohd. Shah Road, Paltan Bazar, Near Starline Hotel, Guwahati-781 008,
Tel.: (0361) 2513881, **Telefax:** 2543669, **E-mail: guwahati@newagepublishers.com**
- **Hyderabad** 105, 1st Floor, Madhiray Kaveri Tower, 3-2-19, Azam Jahi Road, Near Kumar Theater, Nimboliadda Kachiguda, Hyderabad-500 027, **Tel.:** (040) 24652456, **Telefax:** 24652457
E-mail: hyderabad@newagepublishers.com
- **Kolkata** RDB Chambers (Formerly Lotus Cinema) 106A, 1st Floor, S N Banerjee Road, Kolkata-700 014
Tel.: (033) 22273773, **Telefax:** 22275247, **E-mail: kolkata@newagepublishers.com**
- **Lucknow** 16-A, Jopling Road, Lucknow-226 001, **Tel.:** (0522) 2209578, 4045297, **Telefax:** 2204098
E-mail: lucknow@newagepublishers.com
- **Mumbai** 142C, Victor House, Ground Floor, N.M. Joshi Marg, Lower Parel, Mumbai-400 013
Tel.: (022) 24927869, **Telefax:** 24915415, **E-mail: mumbai@newagepublishers.com**
- **New Delhi** 22, Golden House, Daryaganj, New Delhi-110 002, **Tel.:** (011) 23262368, 23262370, **Telefax:** 43551305
E-mail: sales@newagepublishers.com

ISBN: 978-81-224-1887-3
C-17-11-10853

Printed in India at Star Print-O-Bind, Delhi.
Typeset at Kalyani Graphics, Delhi.

NEW AGE INTERNATIONAL (P) LIMITED, PUBLISHERS
7/30 A, Daryaganj, New Delhi-110002
Visit us at **www.newagepublishers.com**
(CIN: U74899DL1966PTC004618)

PREFACE

The microscopic description of a many-particle system requires specification of positions and velocities of all its constituent particles at one instant of time. For example, one mole of a helium gas contains 6.02×10^{23} atoms. To specify the position and velocity of each atom we need six numbers, three position coordinates and three velocity coordinates. So we need $6 \times 6 \times 10^{23}$ numbers for the description of the gas. If an instrument can register one million (10^6) coordinates per second, it would take $6 \times 6 \times 10^{17}$ sec or 12×10^{10} years of time to handle these data. This time is enormously large in comparison to human life. Naturally, this form of information about the individual particle is not suitable. For the investigation of a many-particle system, the form of information must be of a general nature pertaining to a large number of particles rather than to individual particles. So we need a different approach. This new approach is statistical method. The laws describing the behaviour of aggregates of a large number of particles, which are analyzed by statistical methods, are called statistical laws. Statistical mechanics, which is based on statistical laws, provides methods for calculation of macroscopic properties of a many-particle system in terms of microscopic structure of the system. A many-particle system can also be investigated without going into details of its internal structure. In this approach we use the concepts and physical quantities pertaining to the system as a whole. For example, the model of an ideal gas in the state of equilibrium is characterized by pressure, volume and temperature. A relation between these quantities is established by experiments and the theory is developed on the basis of some general postulates. Such a theory is phenomenological in nature and it does not deal with the internal mechanism of the process. Thermodynamics investigates the behaviour of many-particle systems in this way. Statistical mechanics and thermodynamic methods of investigating many-particle systems are complementary to each other. These two approaches help in solving scientific problems in most effective way. So the laws of statistical mechanics and thermodynamics form one of the most important branches of physics. In this introductory book, which is meant of beginners, every attempt has been made to present conceptually most difficult subject in most elementary form. In order to keep the volume of this book within the reasonable limits and to preserve its introductory nature more advanced topics have been omitted.

Inspite of best effort to avoid errors and omissions, some might have crept into the book. I will be grateful to anyone who finds and reports mistakes, deficiencies and misinterpretation in the book.

—R.B. Singh

ABOUT THE PRIMER SERIES

To improve the quality of physics education at university level, U.G.C has recommended a physics course at undergraduate level. Many universities and colleges have adopted almost the same course with minor changes. Teachers and books have to play a vital role in this effort. A book cannot be a substitute of a teacher. It can merely supplement the class-room activities. The action of a teacher such as hand waving, black board sketching, informal discussion and other activities stimulate students and make the learning process more interesting. This does not lessen the importance of good books. While teaching quantum mechanics, statistical mechanics, atomic and molecular physics, semiconductor physics, nuclear physics to undergraduate students, majority of teachers experience great difficulty in finding suitable books to recommend to students. Although there are many excellent books on these topics but their contents are too large and their materials are so arranged that most of the students find themselves in uncomfortable position in selecting sections of their interest. Also a large fraction of students cannot afford these books owing to their high prices. From this situation grew the idea to write a series of introductory books which can cater the needs of the students. Our proposed plan includes following titles:

(*i*) A Primer of Quantum Mechanics

(*ii*) A Primer of Statistical Mechanics

(*iii*) A Primer of Atomic and Molecular Spectra

(*iv*) A Primer of Semiconductor Physics

(*v*) A Primer of Solid State Physics

(*vi*) A Primer of Nuclear Physics

Our aim is to provide materials in the books of this series in such a manner that each title be self-contained and can be used independently of the others.

CONTENTS

1

Preliminary Concepts

1.1 INTRODUCTION

The main objective of statistical mechanics is to predict the properties of a macroscopic system from the knowledge of the behavior of particles constituting the system. In a physical system containing a very large number of particles (atoms and molecules or other constituents) it is usually impossible, for practical reasons, to apply the basic physical laws (classical or quantum) directly to each particle. Instead, it is often advantageous to take a statistical approach, in which one describes the distribution of particles in various states in a statistical manner. The existence of a very large number of particles of the system can be used to advantage in the statistical description. The theory of random processes and quantities form the mathematical tools for this approach.

In statistical mechanics the description of a state of a many particle system is given by stating how the particles are distributed in various allowed microstates. Depending on the nature of the particles, three kinds of statistics or distribution laws are used to describe the properties of the system. The three statistics are:

(1) *Maxwell-Boltzmann or classical statistics*

(2) *Bose-Einstein statistics*

(3) *Fermi-Dirac statistics*

Fremi-Dirac statistics and Bose-Einstein statistics are quantum statistics.

1.2 MAXWELL-BOLTZMANN (M-B) STATISTICS

M-B statistics is applicable to the system of *identical, distinguishable* particles. The particles are so far apart that they are *distinguishable* by their position. In the language of quantum mechanics, the application of classical statistics is valid if the average separation between particles is much greater than the average de Broglie wavelength of the particle. In this situation the wave functions of the particles don't overlap. The particle may have any spin. The classical statistics put no restriction on the number of particles that occupy a state of the system. *M-B* statistics can be safely applied to dilute gases at room and higher temperature.

1.3 BOSE-EINSTEIN (B-E) STATISTICS

B-E statistics is applicable to the system of *identical, indistinguishable* particles, which have *integral spin* (0, 1, 2....). Particles with integral spin are called *bosons*. Bosons don't obey Pauli's exclusion principle. So any number of bosons can occupy a single quantum state. The particles are close enough so that their wave functions overlap. Examples of bosons are *photons(spin 1), phonons (quantum of acoustical vibration), pions, alpha particles, helium atoms* etc.

1.4 FERMI-DIRAC (F-D) STATISTICS

F-D statistics is applicable to the system of *identical, indistinguishable* particles, which have *odd-half-integral spin* (1/2, 3/2,5/2...). Particles with odd-half-integral spin are called *fermions* and they obey Pauli's exclusion principle. Hence, not more than one *fermion* can occupy a quantum state. The *F-D* statistics is valid if the average separation between fermions is comparable to the average de Broglie wavelength of fermions so that their wave functions overlap. Examples of fermions are electrons, positrons, μ-mesons, protons, neutrons etc.

In the limit of high temperature and low particle density the two quantum statistics (*B-E* and *F-D*) yield results identical to those obtained using the classical statistics.

1.5 SPECIFICATION OF THE STATE OF A SYSTEM

A system consisting of micro-particles (such as atoms and molecules) is described by the laws of quantum mechanics. In quantum mechanical description the most precise possible measurement on a system always shows this system to be in some one of a set of discrete quantum states characteristic of the system. The microscopic state of a system can thus be described completely by specifying the particular quantum state in which the system is found. Each quantum state of an isolated system is associated with a definite value of energy and is called an energy level. There may be several quantum states corresponding to the same energy of the system. These quantum states are then said to be degenerate. Every system has a lowest possible energy. There is usually only one possible quantum state of the system corresponding to this lowest energy; this state is said to be the ground state of the system. (Exceptions may be there.)

For illustration we take an example. Consider a particle of mass m restricted to move inside a box of sides L_x, L_y, L_z located at the origin of Cartesian axes such that $0 \le x \le L_x$, $0 \le y \le L_y$, $0 \le z \le L_z$. Schrodinger equation for the particle is

$$\nabla^2\psi + \frac{2mE}{\hbar^2}\psi = 0$$

$$\frac{\partial^2\psi}{\partial x^2} + \frac{\partial^2\psi}{\partial y^2} + \frac{\partial^2\psi}{\partial z^2} + k^2\psi = 0, \quad k^2 = \frac{2mE}{\hbar^2} \tag{1.5.1}$$

The solution of the Eq. (1.5.1) subject to the boundary conditions: $\psi = 0$ at $x = 0$, $x = L_x$, $y = 0$, $y = L_y$, $z = 0$, $z = L_z$ is found to be

$$\psi(x, y, z) = A \sin\frac{n_x \pi x}{L_x} \sin\frac{n_y \pi y}{L_y} \sin\frac{n_z \pi z}{L_z} \tag{1.5.2}$$

where n_x, n_y, n_z are positive integers and each can take on values 1, 2, 3,..... The allowed energies of the particle come out to be

$$E = \frac{\pi^2\hbar^2}{2m}\left[\frac{n_x^2}{L_x^2} + \frac{n_y^2}{L_y^2} + \frac{n_z^2}{L_z^2}\right] \tag{1.5.3}$$

'If $L_x = L_y = L_z = L$ then the energy of the particle is given by

$$E_{n_x n_y n_z} = \frac{\pi^2 \hbar^2}{2mL^2}\left(n_x^2 + n_y^2 + n_z^2\right) \quad (1.5.4)$$

and the state of the particle is given by the wave function

$$\psi(x,y,z) = A \sin\frac{n_x \pi x}{L} \sin\frac{n_y \pi y}{L} \sin\frac{n_z \pi z}{L} \quad (1.5.5)$$

The triad n_x, n_y, n_z defines a quantum state of the particle. For ground state $n_x = n_y = n_z = 1$. This state is represented as $\psi_{111}(x, y, z)$. The energy in this state is $E_{111} = \frac{3\pi^2\hbar^2}{2mL^2}$. A single quantum state corresponds to the lowest energy level. The ground state is thus non-degenerate. The degeneracy of an energy level is given by the number of ways that the integer $(n_x^2 + n_y^2 + n_z^2)$ can be written as the sum of squares of the three positive integers. A few energy levels are given below.

$$E_{111} = \frac{3\pi^2\hbar^2}{2mL^2}, E_{112} = E_{121} = E_{211} = \frac{6\pi^2\hbar^2}{2mL^2},$$

$$E_{122} = E_{212} = E_{221} = \frac{9\pi^2\hbar^2}{2mL^2}$$

$$E_{113} = E_{131} = E_{311} = \frac{11\pi^2\hbar^2}{2mL^2},$$

$$E_{222} = \frac{12\pi^2\hbar^2}{2mL^2},$$

$$E_{123} = E_{132} = E_{213} = E_{231} = E_{321} = E_{312} = \frac{14\pi^2\hbar^2}{2mL^2},$$

$$E_{223} = E_{232} = E_{322} = \frac{17\pi^2\hbar^2}{2mL^2}$$

Notice that the second, third and fourth energy levels are 3-fold degenerate, the fifth energy level is non-degenerate, sixth energy level is 6-fold degenerate and the seventh energy level is 3-fold degenerate and so on.

		Degeneracy
$E_{223}, E_{132}, E_{322}$	————	3
E_{123}, E_{132}, E_{213}, E_{231}, E_{321}, E_{312}	————	6
E_{222}	————	1
E_{113}, E_{131}, E_{311}	————	3
E_{122}, E_{212}, E_{221}	————	3
E_{112}, E_{121}, E_{211}	————	3
E_{111}	————	1

Fig. 1.5.1 Energy levels of a particle in a box

1.6 DENSITY OF STATES

The allowed energy levels and associated quantum states for a particle confined to move in a cubical enclosure of side L are given by

$$E_{n_1n_2n_3} = \frac{p^2}{2m} = \frac{\hbar^2k^2}{2m} = \frac{\hbar^2}{2m}\left(k_1^2 + k_2^2 + k_3^2\right)$$

$$= \frac{\hbar^2}{2m}\left[\left(\frac{n_1\pi}{L}\right)^2 + \left(\frac{n_2\pi}{L}\right)^2 + \left(\frac{n_3\pi}{L}\right)^2\right] \tag{1.6.1}$$

$$\psi(x,y,z)_{n_1n_2n_3} = const.\sin\frac{n_1\pi\, x}{L}\sin\frac{n_2\pi\, y}{L}\sin\frac{n_3\pi\, z}{L} \tag{1.6.2}$$

where n_1, n_2 and n_3 are non-zero positive integers. The wave function or the state describing the particle has wave vector $k = \left\{\frac{n_1\pi}{L}, \frac{n_2\pi}{L}, \frac{n_3\pi}{L}\right\}$.

We can plot the components of the wave vector $\mathbf{k}$ in three-dimensional space with k_1, k_2, k_3 as Cartesian axes. This space is called k-space. In k-space the allowed values $\boldsymbol{k}$ form a cubical point lattice with spacing between points being π/L. Each lattice point in k-space represents a permissible state of the particle. These lattice points divide the k-space into cells, each of volume $(\pi/L)^3$. The contribution to the unit cell of points lying at the corners of the unit cell is unity. Each lattice point, which corresponds to a quantum states, occupies a volume $(\pi/L)^3$ in k-space.

We wish to find the number of quantum states with wave vectors whose magnitude lie in the interval k and $k + dk$. This number is equal to the number of lattice points in k-space lying between two spherical shells, centered at the origin, of radii k and $k + dk$ in the positive octant. The volume of the region lying between the radii k and $k + dk$ in the positive octant is $\frac{1}{8}(4\pi k^2\, dk)$. So the number of states with wave vectors whose magnitudes lie in the range k to $k + dk$ is

$$g(k)dk = \frac{\frac{1}{8}(4\pi k^2 dk)}{(\pi / L)^3} = \frac{V}{2\pi^2}k^2 dk \tag{1.6.3}$$

where $V = L^3$ is the volume of the enclosure. The function g (k), which represents the number of quantum states per unit energy range at energy E, is called the *density of states*.

Making use of the relations

$$p = \hbar k = \sqrt{2mE}$$

the expression for the density of states can be written as

$$g(p)dp = \frac{V}{h^3}4\pi\, p^2 dp \tag{1.6.4}$$

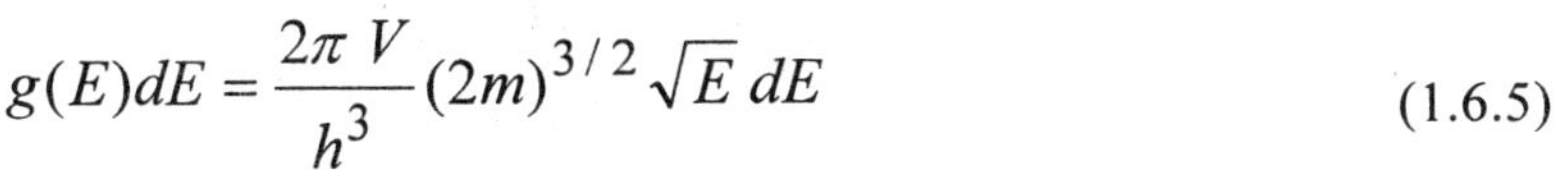

$$g(E)dE = \frac{2\pi V}{h^3}(2m)^{3/2}\sqrt{E}\, dE \tag{1.6.5}$$

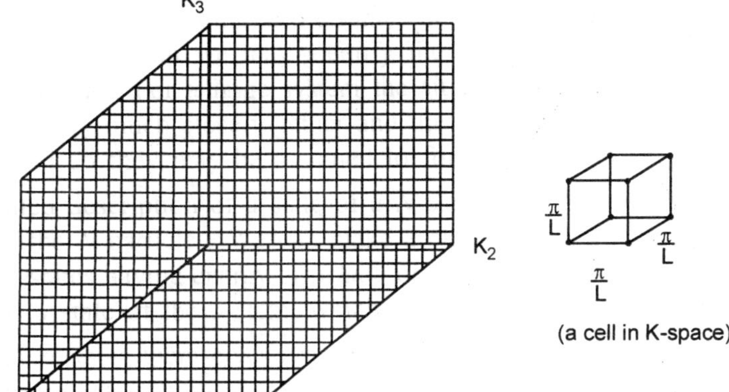

Fig. 1.6.1 Representation of states in *k*-space.

Periodic Boundary Conditions

The formula for the density of states is independent of the detailed form of boundary conditions imposed at the surface of the enclosure. We shall show this by using an alternative boundary condition, the periodic boundary condition, which is most often used. For a cubic enclosure of side L this condition is expressed as

$$\begin{aligned} \psi(0,y,z) &= \psi(L,y,z) \\ \psi(x,\upsilon,z) &= \psi(x,L,z) \\ \psi(x,y,0) &= \psi(x,y,L) \end{aligned} \tag{1.6.6}$$

The solution of Schrodinger wave equation for a particle in a box is

$$\psi(x,y,z) = const.\ \exp i\{k_1x + k_2y + k_3z) \tag{1.6.7}$$

The wave vector **k** is now restricted to the values

$$k = \left\{\frac{2\pi}{L}n_1, \frac{2\pi}{L}n_2, \frac{2\pi}{L}n_3\right\} \tag{1.6.8}$$

Notice that the n_i's now can be positive or negative integers. Now, to calculate the density of states, instead of taking positive octant of a sphere in k-space, we take the whole sphere. The spacing of lattice points in k-space is now $2\pi/L$. The number of states whose wave vector $\boldsymbol{k}$ has magnitude in the range k and $k + dk$ is given by

$$g(k)dk = \frac{4\pi k^2 dk}{(2\pi/L)^3} = \frac{V}{2\pi}k^2 dk \tag{1.6.9}$$

If we take $E = (3/2)\ kT$, $T = 300\ K$, $m = 10^{-22}$ g, L = 10 cm, $dE = 0.01\ E$, then

$$g(E)dE \cong 10^{28}$$

So even for a system as simple as a particle in a box, the number of states can be very large at room temperature.

For an N-particle system, the degeneracy is tremendously large. To see this, consider a system of N non-interacting particles in a cube of side L. The energy of the system is

$$E = \frac{\pi^2 \hbar^2}{2mL^2} \sum_{i=1}^{N} \left[n_{xi}^2 + n_{yi}^2 + n_{zi}^2 \right] \tag{1.6.10}$$

Using the volume of an N-dimensional sphere the number of states with energy $\leq E$ is

$$\Phi(E) = \frac{1}{\Gamma(n+1)\,\Gamma\left(\frac{3N}{2}+1\right)} \left(\frac{2\pi m L^2 E}{4\pi^2 h^2} \right)^{3N/2}$$

and

$$g(E)dE = \frac{1}{\Gamma(N+1)\,\Gamma(N/2)} \left(\frac{2\pi m L^2}{4\pi^2 h^2} \right)^{3N/2} E\left(\frac{3N}{2} - 1 \right) dE \tag{1.6.11}$$

If $E = (3/2)kT$, $T = 300K$, $m = 10^{-22}$ g, $L = 10$ cm, $N = 6.02 \times 10^{23}$, $\Delta E = 0.01E$ we have $g(E)dE = 10^N$

This shows that as the number of particles in the system increases, the quantum mechanical degeneracy becomes enormous. For a system consisting of 10^{23} particles the allowed states are so croweded that it is impossible to enumerate and work with individual states. The best we can do is to work with density of states $g(E)$, which is the number of states per unit energy range. For a large system the density of states may be taken to be a smooth function of energy. The density of states rises very rapidly with energy in a large system.

1.7 MACROSCOPIC (MACRO) STATE

Consider a system containing a very large number N of particles in a vessel of fixed volume V. The total energy of the system is E. The state of the system specified by its pressure P, temperature T, volume V and energy E is called the macroscopic state of the system. If the system is in equilibrium, the macroscopic characteristics P, V, T, E don't vary with time.

1.8 MICROSCOPIC (MICRO) STATE

The most complete description of many-particle system is given by specifying the positions and momenta of its constituent particles. The state of the system characterized by positions and momenta of all its particles is called the microscopic or microstate. The state of a particle moving in space is specified by 3 spatial coordinates (x, y, z) and 3 momentum coordinates (p_x, p_y, p_z). These 6 numbers x, y, z, p_x, p_y, p_z completely determine the state of the particle. If there are N particles in the system, the state of the entire system is specified by $6N$ numbers: of which $3N$ are spatial coordinates and $3N$ are momentum coordinates. In equilibrium the macroscopic variables P, V, T characterizing the system are independent of time. However, the particles of the system are in random motion and the microscopic states of the system undergo continuous change in course of time. So there are an enormously large number of microscopic states corresponding to each macrostate. In other words a macrostate is realized through an enormously large number of microstates. The aim of the statistical mechanics is to establish a relation between macrostate and microstates.

The statistical treatment of a thermodynamic system may be developed using either classical or quantum mechanics. In what follows we shall use quantum mechanics at most points. Objects of real world obey quantum mechanics. Objects described by quantum mechanics don't usually have arbitrary internal energy. Bounded systems exist only in certain well-defined energy and well-defined quantum states. In other words the energy of a system restricted to certain region of space is quantized. The allowed energies of the system are called energy levels. In such systems one also finds that some distinct states have the same energy, such states are said to be **degenerate**. The number of distinct states corresponding to the same energy level is called the degree of degeneracy of the level. For a small system we may identify the various quantum states and their energies without too much difficulty. For a large system the situation is quite different. The energy levels of a large system are very much close together and the mean separation between the energy levels is extremely small and so they may be assumed to form continuum.

The most detailed description of a state of a N-particle system is given by a specification of the state of each of the N particles. We can make a chart showing *which* particles are in each of the various quantum states having energy ε_1, *which* ones in the states with energy ε_2 and so on. This description specifies a state of the system, which we call a microscopic state.

Usually, the microstates themselves are not very useful. If the particles are identical, it is of no concern precisely *which* particles are in *which* energy states. Instead, we want to know *how many* particles are in a given energy state, without regard to *which* particle they are. Thus a different and more directly useful kind of the state description consists in specifying only the number of particles in each of the possible energy level.

The specification that there are

n_1 particles in energy level ε_1 with degeneracy g_1

n_2 particles in energy level ε_2 with degeneracy g_2

....

n_i particles in energy level ε_i with degeneracy g_i

is a description of a *macrostate* of the system. The number of ways Ω in which this macrostate may be achieved is called microstates of that macrostate. The quantity Ω is called *statistical weight or thermodynamic probability* of that macrostate. The larger Ω is, the greater the probability of finding the system in that macrostate. If the volume V, the number of particles N and the total energy E of the system is kept constant, the equilibrium state of the system will correspond to that macrostate in which Ω is maximum. The principal objective of statistical mechanics is to determine the possible distribution of particles among the various energy levels and quantum states. If the distribution of particles of a system among its quantum states is known, the macroscopic properties of the system can be determined.

If a system is composed of N identical and ***distinguishable*** particles, the total number of microstates Ω corresponding to a macrostate specified by the numbers $\{n_1, n_2, \ldots\}$ is given by

$$\Omega = \frac{N!}{n_1!n_2!\ldots\ldots}(g_1)^{n_1}(g_2)^{n_2}\ldots\ldots\ldots \quad (1.8.1)$$

In a gas containing N molecules, the molecules are ***distinguishable*** if the mean separation between the molecules is much larger than their de Broglie wavelength. In deriving the above formula for the number of microstates, it is assumed that there is no restriction on the number of particles that can occupy a quantum state.

If the system is composed of N ***bosons***, the total number of microstates corresponding to a macrostate specified by the numbers $\{n_1, n_2, \ldots\}$ is given by

$$\Omega = \frac{(n_1 + g_1 - 1)!}{n_1!(g_1 - 1)!} \cdot \frac{(n_2 + g_2 - 1)!}{n_2!(g_2 - 1)!} \cdots\cdots\cdots = \prod_i \frac{(n_i + g_i - 1)!}{n_i!(g_i - 1)!} \tag{1.8.2}$$

If the system is composed of *N* ***fermions***, the total number of microstates corresponding to a macrostate specified by the numbers $\{n_1, n_2, \ldots\}$ is given by

$$\Omega = \frac{g_1!}{n_1!(g_1 - n_1)!} \cdot \frac{g_2!}{n_2!(g_2 - n_2)!} \cdots\cdots = \prod_i \frac{g_i!}{n_i!\,(g_i - n_i)!} \tag{1.8.3}$$

Example 1: *Two particles are to be distributed in an energy level, which is 3 fold-degenerate. Find the possible microstates if the particles are (i) distinguishable, (ii) indistinguishable bosons, (iii) indistinguishable fermions.*

Sol. (*i*) If the particles are distinguishable, they can be labeled as *A* and *B*. Here $N = 2$, $n_1 = 2$, $g_1 = 3$. The number of possible microstates is:

$$\Omega = \frac{N!}{n_1!n_2!\ldots\ldots}(g_1)^{n_1}(g_2)^{n_2}\ldots\ldots$$

$$\Omega = \frac{2!}{2!}(3)^2 = 9$$

A	B	
B	A	
AB		

	A	B
	B	A
	AB	

A		B
B		A
		AB

It is worth to note that if there are more than one particle in a given quantum state, an interchange of order in which the labeled objects appear does not produce a new microstate. So the *AB* and *BA* are the same.

(*ii*) If the particles are indistinguishable **bosons**, the number of possible microstates is

$$\Omega = \frac{(n_1 + g_1 - 1)!}{n_1!(g_1 - 1)!} = \frac{4!}{2!2!} = 6$$

•	•	
	•	•
•		•
••		
	••	
		••

The number of microstates is 6.

(*iii*) The particles are indistinguishable **Fermions.**

The number of microstates is $\Omega = \dfrac{g_1!}{n_1(g_1 - n_1)!} = \dfrac{3!}{2!1!} = 3.$

•	•	

	•	•

•		•

Example 2. *Four particles are to be distributed among four energy levels $\varepsilon_1 = 1$, $\varepsilon_2 = 2$, $\varepsilon_3 = 3$, $\varepsilon_4 = 4$ units having degeneracies $g_1 = 1$, $g_2 = 2$, $g_3 = 2$, $g_4 = 1$ respectively. The total energy of the system is 10 units. Find the possible distribution (macrostates) and the microstates corresponding to most probable macrostate. Assume that the particles are (i) distinguishable, (ii) indistinguishable bosons, (iii) indistinguishable fermions.*

Sol. (*i*) The possible macrostates are

$\Omega_{2,0,0,2} = \{2,0,0,2\}$, $\Omega_{1,1,1,1} = \{1,1,1,1\}$, $\Omega_{0,3,0,1} = \{0,3,0,1\}$, $\Omega_{1,0,3,0} = \{1,0,3,0\}$, $\Omega_{0,2,2,0} = \{0,2,2,0\}$.

$\varepsilon_4 = 4$	o o	o	o	———	———
$\varepsilon_3 = 3$	———	o	———	o o o	o o
$\varepsilon_2 = 2$	———	o	o o o	———	o o
$\varepsilon_1 = 1$	o o	o	———	o	———

The number of microstates in above macrostates is

$$\Omega_{2,0,0,2} = \frac{4!}{2!0!0!2!} = 6$$

$$\Omega_{1,1,1,1} = \frac{4!}{1!1!1!1!} = 24$$

$$\Omega_{0,3,0,1} = \frac{4!}{0!3!0!1!} = 4$$

$$\Omega_{1,0,3,0} = \frac{4!}{1!0!3!0!} = 4$$

$$\Omega_{0,2,2,0} = \frac{4!}{0!2!2!0!} = 6$$

The most probable macrostate is {1, 1, 1, 1}. The possible microstates corresponding to this macrostate are shown in the table. Since the particles are distinguishable, they have been labeled *A*, *B*, *C* and *D*.

The number of microstates corresponding to the macrostate {1,1,1,1} is:

$$\Omega_{1,1,1,1} = \frac{N!}{n_1!n_2!....}(g_1)^{n_1}(g_2)^{n_2} ..... = \frac{4!}{1!1!1!1!}(1)^1(2)^1(2)^1(1)^1 = 96$$

Here $n_1 = 1$, $n_2 = 1$, $n_3 = 1$, and $n_4 = 1$, and $g_1 = 1$, $g_2 = 2$, $g_3 = 2$, $g_4 = 1$.

	1	2	3	4	5	6	7	8	9	10	11	12	13	14	15	16	17	18	19	20	21	22	23	24
ε_4	A	A	A	A	A	A	B	B	B	B	B	B	C	C	C	C	C	C	D	D	D	D	D	D
ε_3	B	B	C	C	D	D	A	A	C	C	D	D	A	A	B	B	D	D	A	A	B	B	C	C
ε_2	C	D	B	D	B	C	C	D	A	D	A	C	B	D	A	D	A	B	B	C	A	C	A	B
ε_1	D	C	D	B	C	B	D	C	D	A	C	A	D	B	D	A	B	A	C	B	C	A	B	A

The macrostate {1,1,1,1} has 24 microstates as shown in the table. If the degeneracies of the second and third levels are considered, it is found that each of the 24 microstates has 4 microstates. The microstates corresponding to the first microstate are shown below.

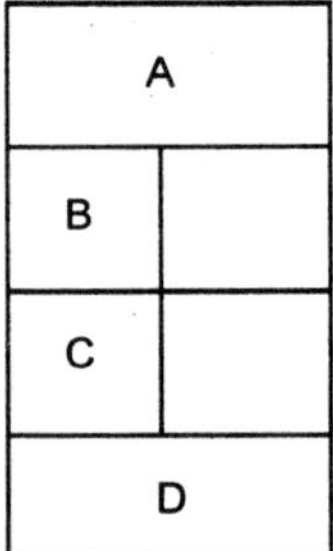

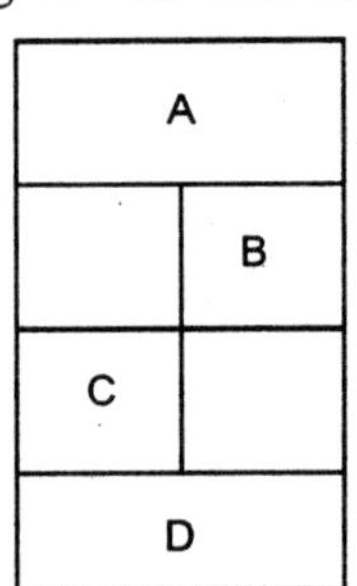

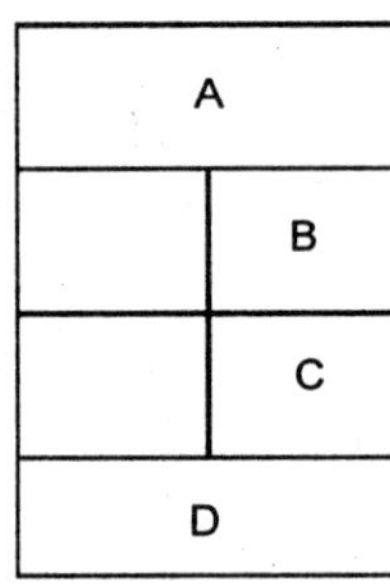

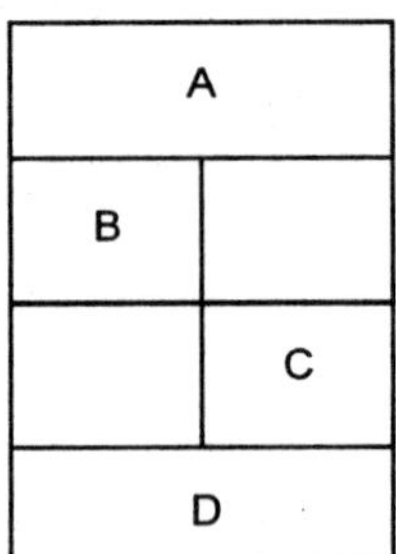

Similarly each of the 24 microstates gives 4 microstates. Thus the total number of microstates corresponding to the macrostate {1,1,1,1} is 96.

(*ii*) Particles are **bosons.**

The possible macrostates are

{2,0,0,2}, {1,1,1,1}, {0,3,0,1}, {1,0,3,0}, {0,2,2,0}

ε_{1i}	g_i	macrostates				
		$\{n_i\}$	$\{n_i\}$	$\{n_i\}$	$\{n_i\}$	$\{n_i\}$
4	1	2	1	1	0	0
3	2	0	1	0	3	2
2	2	0	1	3	0	2
1	1	2	1	0	1	0

The number of microstates associated with a macrostate $\{n_1, n_2, ...\}$ is

$$\Omega = \prod_i \frac{(n_i + g_i - 1)!}{n_i!(g_i - 1)!}$$

$$\Omega_{2,0,0,2} = \frac{(2+1-1)!}{2!(1-1)!}\frac{(0+2-1)!}{0!(2-1)!}\frac{(0+2-1)!}{0!(2-1)!}\frac{(2+1-1)!}{2!(1-1)!} = 1$$

Similarly $\Omega_{1,1,1,1} = 4$, $\Omega_{0,3,0,1} = 4$, $\Omega_{1,0,3,0} = 4$, $\Omega_{0,2,2,0} = 9$

The most probable macrostate is {0,2,2,0}. The various microstates associated with this macrostate are shown in the figure.

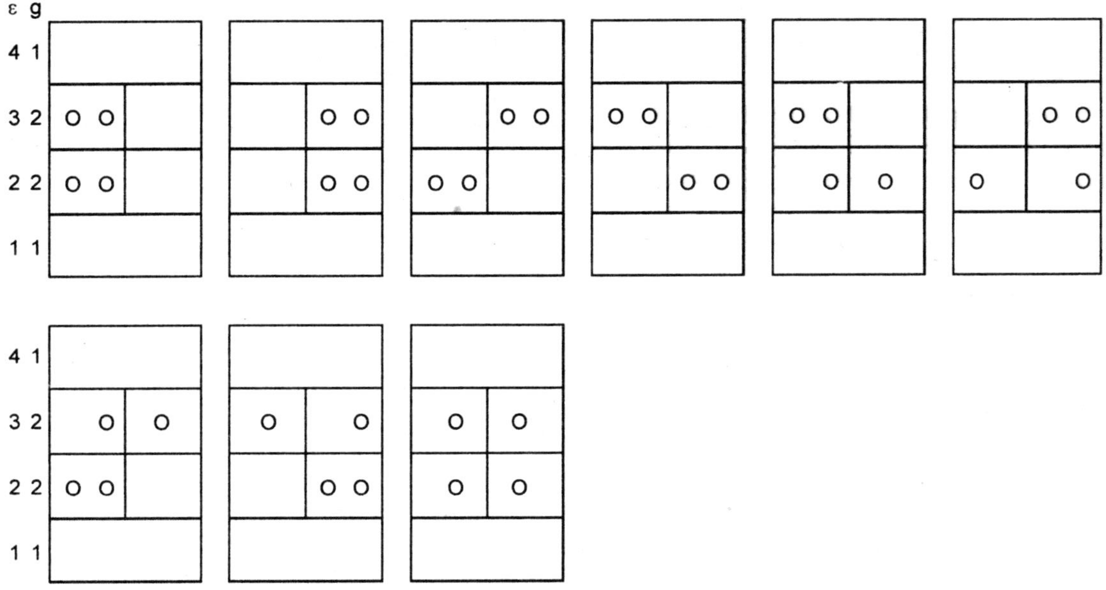

(*iii*) Particles are **fermions.**

There are two possible macrostates. They are {1,1,1,1} and {0,2,2,0}. The number of possible microstates associated with the first macrostate is

$$\Omega_{1,1,1,1} = \Pi \frac{g_i!}{n_i!(g_i - n_i)!} = \frac{1!}{1!(1-1)!} \frac{2!}{1!(2-1)!} \frac{2!}{1!(2-1)!} \frac{1!}{1!(1-1)!} = 4$$

Similarly, the number of microstates associated with the second macrostate is

$$\Omega_{0,2,2,0} = 1$$

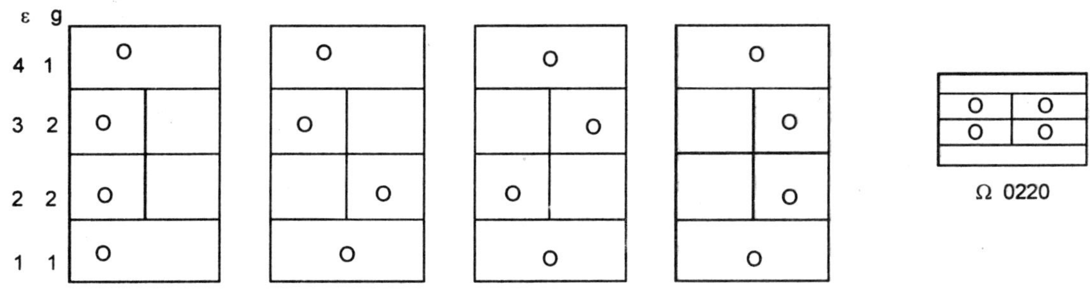

(four microstates corresponding to macrostate $\Omega_{1,1,1,1}$

(one microstate)

Example 3: *A system consists of 4 distinguishable particles, labeled 1, 2, 3, and 4. These particles are to be distributed in two non-degenerate energy levels ε_1 and ε_2. Find the possible macrostates and corresponding microstates. There is no restriction on the number of particles that can be put in a quantum state.*

Sol. For non-degenerate levels $g_1 = g_2 \ldots\ldots = 1$. So the number of microstates corresponding to a macrostate is given by

$$\Omega = \frac{N!}{n_1!n_2!\ldots}$$

The possible macrostates are {4, 0}, {3, 1}, {2, 2}, {1, 3} and {0, 4}. The number of microstates corresponding to these macrostates is:

$$\Omega_{4,0} = \frac{4!}{4!0!} = 1,\ \Omega_{3,1} = \frac{4!}{3!1!} = 4,\ \Omega_{2,2} = \frac{4!}{2!2!} = 6,\ \Omega_{1,3} = \frac{4!}{1!3!} = 4,\ \Omega_{0,4} = \frac{4!}{0!4!} = 1$$

So there are five macrostates and 16 microstates in total. The macrostate $\Omega_{2,2}$ has maximum number of microstates (=6) and hence it is the most probable macrostate of the system. These microstates are shown in the table.

Energy Levels ε_1	ε_2	*Macrostate* $\{n_1, n_2\}$	*Microstates*
1 2 3 4	X	{4, 0}	1
1 2 3	4		
1 2 4	3	{3, 1}	4
1 3 4	2		
2 3 4	1		
1 2	3 4		
1 3	2 4		
1 4	2 3	{2, 2}	6
2 3	1 4		
2 4	1 3		
3 4	1 2		
1	2 3 4		
2	1 3 4	{1, 3}	4
3	1 2 4		
4	1 2 3		
X	1 2 3 4	{0, 4}	1

Example 4: *A system consisting of 4 particles has total energy 12 units. The particles are to be distributed in 4 non-degenerate energy levels with energies $\varepsilon_1 = 1$, $\varepsilon_2 = 2$, $\varepsilon_3 = 3$ and $\varepsilon_4 = 4$ units . Find the possible macrostates and their microstates. Assume that any number of particles can be put in the allowed energy levels.*

Sol. The possible macrostates are:

{1, 0, 1, 2}, {0, 2, 0, 2}, {0, 1, 2, 1}, {0, 0, 4, 0}

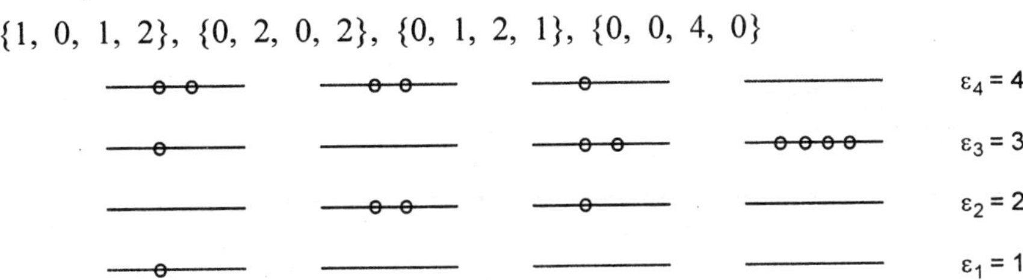

The number of microstates Ω belonging to the above macrostates is:

$$\Omega_{1,0,1,2} = \frac{4!}{1!0!1!2!} = 12$$

$$\Omega_{0,2,0,2} = \frac{4!}{0!2!0!2!} = 6$$

$$\Omega_{0,1,2,1} = \frac{4!}{0!1!2!1!} = 12$$

$$\Omega_{0,0,4,0} = \frac{4!}{0!0!4!0!} = 1$$

Example 5: *A system composed of 6 bosons has total energy 6 units. These particles are to be distributed in energy levels $\varepsilon_0 = 0$, $\varepsilon_1 = 1$, $\varepsilon_2 = 2$, $\varepsilon_3 = 3$, $\varepsilon_4 = 4$, $\varepsilon_5 = 5$, $\varepsilon_6 = 6$. Each level is triply degenerate. Calculate the thermodynamic probability (statistical weight) of all the macrostates and microstates of the system.*

Sol. The possible macrostates are {5,0,0,0,0,0,1}, {4,1,0,0,0,1,0}, {4,0,1,0,1,0,0}, {3,2,0,0,1,0,0,}, {4,0,0,2,0,0,0}, {3,1,1,1,0,0,0}, {2,3,0,1,0,0,0}, {3,0,3,0,0,0,0}, {2,2,2,0,0,0,0}, {1,4,1,0,0,0,0}, {0,6,0,0,0,0,0}. There are 11 macrostates.

The thermodynamic probability (statistical weight) of a macrostate is given by

$$\Omega = \prod_i \frac{(n_i + g_i - 1)!}{n_i!(g_i - 1)!}$$

$\Omega_I = 63$, $\Omega_{II} = 135$, $\Omega_{III} = 135$, $\Omega_{IV} = 180$, $\Omega_V = 90$, $\Omega_{VI} = 270$, $\Omega_{VII} = 180$, $\Omega_{VIII} = 100$, $\Omega_{IX} = 216$, $\Omega_X = 135$, $\Omega_{XI} = 28$,

The thermodynamic probability of the system is

$$\Omega = \sum_i \Omega_i = 63 + 135 + 135 + 180 + 90 + 270 + 180 + 100 + 216 + 135 + 28 = 1532$$

Note: For fermions, the macrostates 1, 2, 3, 5, 10 and 11, in which there can be more than one particle in a quantum state, are not allowed.

	1	2	3	4	5	6	7	8	9	10	11
ε_6	o										
ε_5		o									
ε_4			o	o							
ε_3					oo	o	o				
ε_2			o			o		ooo	oo	o	
ε_1		o		oo		o	ooo		oo	oooo	oooooo
ε_0	ooooo	oooo	oooo	ooo	oooo	ooo	oo	ooo	oo	o	

Example 6: *A system consisting of 6 fermions has total energy 6 units. These particles are to be distributed in 5 energy levels $\varepsilon_0 = 0$, $\varepsilon_1 = 1$, $\varepsilon_2 = 2$, $\varepsilon_3 = 3$, $\varepsilon_4 = 4$ units. Each energy level is triply degenerate. Find the possible microstates.*

Sol. The possible macrostates are 5. They are:

{3,2,0,0,1}, {3,1,1,1,0}, {2,3,0,1,0}, {3,0,3,0,0}, {2,2,2,0,0}

The thermodynamic probability (the number of microstates) of a macrostate is given by

$$\Omega = \prod_i \frac{g_i!}{n_i!(g_i - n_i)!}$$

The number of microstates associated with first macrostate is

$$\Omega_I = \frac{3!}{3!(3-3)!} \cdot \frac{3!}{2!(3-2)!} \cdot \frac{3!}{0!(3-0)!} \cdot \frac{3!}{0!(3-0)!} \cdot \frac{3!}{3!(3-1)!} = 9$$

Similarly, the number of microstates associated with other macrostates can be calculated. They come out to be

$$\Omega_{II} = 27, \quad \Omega_{III} = 9, \quad \Omega_{IV} = 1, \quad \Omega_V = 27.$$

The thermodynamic probability of the system is $\Omega = \Sigma\, \Omega_i = 9 + 27 + 9 + 1 + 27 = 73$

2

Phase Space

2.1 PHASE SPACE

The specification of the state of a particle in classical mechanics involves the concept of phase space. To understand the meaning of phase space, consider a particle moving in one dimension, along the x-axis, say. In classical mechanics the state of motion at any instant is specified by specifying its position coordinate x and momentum coordinate p_x. Now imagine a two dimensional conceptual space with x and p_x as orthogonal axes. We call this space phase space. At any instant t, the state of the particle is represented by a point (x, p_x) in the phase space and this point is called *phase point* or *representative point*. As the particle moves on the straight line, x and p_x take on different values and the corresponding representative point traces a trajectory in phase space. Thus the evolution of successive states of the particle will be represented by a trajectory in phase space. A point on the trajectory in phase space represents a definite state of motion of the particle.

Usually the spatial coordinate of a particle is denoted by q and the corresponding momentum by p. Then the state of the particle is specified by stating its position and momentum coordinates (q, p). To specify the state of the particle more precisely, it is convenient to subdivide the ranges of the variables q and p into arbitrarily small discrete interval. One can choose fixed intervals of size δq for the subdivision of q and fixed interval of size δp for the subdivision of p. The phase space is then subdivided into small cells of equal size and of two dimensional volume (*i.e.*, area) $\delta q \delta p = h_0$ where h_0 is some small constant having the dimensions of angular momentum. The state of the system can then be specified by stating that its coordinates lies in some interval between q and $q + \delta q$ and its momentum lies in some interval between p and $p + \delta p$ *i.e.*, by stating that the representing point (q, p) lies in a particular cell of phase space. The specification of the state of the system clearly becomes more precise as one decreases the size of the cells into which phase space has been divided *i.e.*, as one decreases the magnitude of h_0. Of course, h_0 can be chosen arbitrarily small in this classical description.

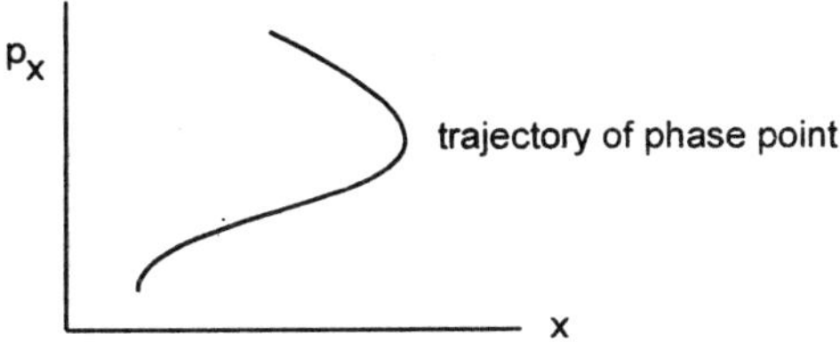

Fig. 2.1.1 Representation of the state of a particle.

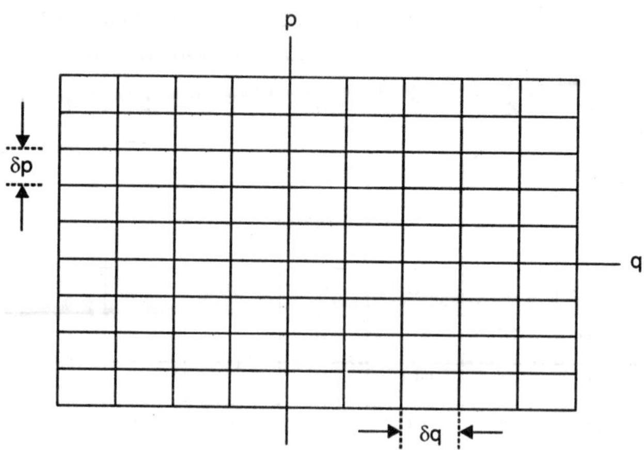

Fig. 2.1.2 Subdivision of phase space in cells of size $\delta q\ \delta p = h_o$.

A particle moving in three dimensions requires 3 position coordinates (x, y, z) and 3 momentum coordinates (p_x, p_y, p_z) to specify its state of motion. The corresponding phase space has 6 dimensions with spatial coordinates x, y, z, and momentum coordinates p_x, p_y, p_z as orthogonal axes and the state of the particle is specified by a point (x, y, z, p_x, p_y, p_z). If we denote the position and momentum coordinates by q_1, q_2, q_3, p_1, p_2, p_3 respectively then the state of the particle is denoted by point (q_1, q_2, q_3, p_1, p_2, p_3). Each point in this 6-dimensional phase space represents a possible state of motion of the particle. If the particle under consideration is a molecule of a gas, the corresponding 6-dimensional phase space is called μ-space. (μ stands for molecule). The state of a system consisting of N particles at any instant will be represented by N phase points in μ-space.

Likewise the state of a N-particle system is specified by $3N$ position coordinates x_1, y_1, z_1; x_2, y_2, z_2; ....x_N, y_N, z_N and $3N$ momentum coordinates p_{x_1}, p_{y_1}, p_{z_1};..........p_{x_N}, p_{y_N}, p_{z_N}. Instead of denoting position coordinates by x, y, z and momentum coordinates by p_x, p_y, p_z let us denote them by generalized position coordinates q_1, q_2, q_3 and by generalized momentum coordinates p_1, p_2, p_3 respectively. So the state of the N-particle system is specified by 3N generalized position coordinates q_1, q_2, q_3 ....q_{3N-2}, q_{3N-1}, q_{3N} and corresponding generalized momentum coordinates p_1, p_2, p_3, ...p_{3N-2}, p_{3N-1}, p_{3N}. These $6N$ coordinates along with the equations of motion viz Hamilton's equations $\dot{q}_k = \dfrac{\partial H}{\partial p_k}$, $\dot{p}_k = -\dfrac{\partial H}{\partial q_k}$, $k = 1,2,3.......3N$, where H is Hamiltonian of the system, completely determine the behaviour of the system. We now imagine a $6N$ dimensional space with $6N$ rectangular axes, one for each of the spatial coordinates q_1........q_{3N} and for each momentum coordinate p_1.........p_{3N}. This $6N$ dimensional phase space of the N-particle system is called γ-space. The state of the entire system (gas) at any time t is completely specified by a phase point in γ–space. In course of time the spatial and momentum coordinates undergo continuous change, the corresponding phase point traces a trajectory in γ-space; the motion of the phase point is governed by Hamilton's equations. Each phase point on the trajectory in γ-space represents a possible microstate of the entire system.

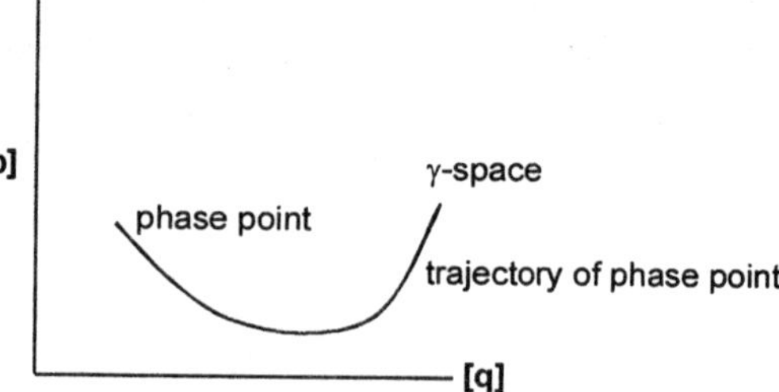

In figure [q] stands for $3N$ spatial coordinates and [p] for $3N$ momentum coordinates.

Fig. 2.13 Trajectory of phase point in γ-space.

Once again the γ-space can be subdivided into little cells of volume $\delta q_1 \ldots . \delta q_{3N}\, \delta p_1 \ldots . \delta p_{3N}$ $= (h_0)^{3N}$. The state of the system can then again be specified by stating in which particular range or cell in phase space, the coordinates $q_1, \ldots . q_{3N}, p_1 \ldots . . p_{3N}$ of the system can be found.

2.2 DENSITY OF STATES IN PHASE SPACE

Consider a particle moving in x-direction. Classical physics puts no restriction on the accuracy with which the simultaneous specification of position and momentum can be made. That is the uncertainties δq and δp in specification of position and momentum can be made as small as we like. In other words the size of the cell $\delta q \delta p = h_0$ into which the phase space is subdivided can be chosen arbitrarily small. But quantum mechanics imposes a limitation on the accuracy with which a simultaneous specification of coordinate q and its corresponding momentum p can be made. According to Heisenberg uncertainty principle, the uncertainties δq and δp are such that $\delta q \delta p \geq h$, where h is Planck's constant. The subdivision of the phase space into cells of volume less than h is physically meaningless.

For a particle free to move in three dimensions, if Δx and Δp_x denote the uncertainty in position and momentum then

$$\Delta x . \Delta p_x \approx h$$

Similar relations hold for other components also.

$$\Delta y . \Delta p_y \approx h$$
$$\Delta z . \Delta p_z \approx h$$

Hence

$$\Delta x . \Delta y . \Delta z .\ \Delta p_x . \Delta p_y . \Delta p_z \gg h^3 \tag{2.1.1}$$

The product $\Delta x . \Delta y . \Delta z .\ \Delta p_x . \Delta p_y . \Delta p_z = h^3$ represents an element of volume in phase space. Two particles whose representative points lie in such an elementary cell cannot be distinguished and hence the representative points in this cell represent a single quantum state. It follows from above that different quantum states shall corresponds to different elements of volume in the phase space only if the size of these elements is no less than h^3. Therefore a volume equal to h^3 in the phase space may be allotted to each microscopic state of the particle. In other words each elementary cell of phase space represents a microstate of the particle. The state of a particle is specified by stating in which particular cell the coordinates x, y, z, p_x, p_y, p_z of the particle lie. The process of dividing the phase space into cells finite size is termed quantization of phase space. The number of quantum states (microstates) in the allowed region of phase space is given by

$$g(x, y, z, p_x, p_y, p_z) dxdydzdp_x dp_y dp_z = \frac{\iiiiiint dxdydzdp_x dp_y dp_z}{h^3} \tag{2.1.2}$$

For a particle of mass m confined to move in a cubical box of side L the number of quantum states is

$$g(p) d^3 p = \frac{\int_0^L \int_0^L \int_0^L dxdydz \iiint dp_x dp_y dp_z}{h^3} = \frac{V}{h^3} \iiint dp_x dp_y dp_z \tag{2.1.3}$$

where $d^3 p = dp_x\, dp_y\, dp_z$ represents a volume element at point (p_x, p_y, p_z) in momentum space. In polar coordinates the volume element in momentum space is given by $p^2 dp \sin\theta\, d\theta\, d\phi$. The number of states such that the momentum of particle lies in the range dp about p irrespective of its directions is given by

$$g(p)dp = \frac{V}{h^3} \int_p^{p+dp} p^2 dp \int_0^{\pi} \sin\theta \int_0^{2\pi} d\varphi$$

$$= \frac{V}{h^3}\left(4\pi p^2 dp\right) \quad (2.1.4)$$

In terms of energy $E = p^2/2m$ the number of states in the energy range dE about E is given by

$$g(E)dE = \frac{2\pi V}{h^3}(2m)^{3/2} E^{1/2} dE \quad (2.1.5)$$

If the particle has internal degree of freedom such as spin, there will be $(2s + 1)$ spin states corresponding to each momentum or energy states. Therefore the number of states then becomes

$$g(E)dE = (2s+1)\frac{2\pi V}{h^3}(2m)^{3/2} E^{1/2} dE \quad (2.1.6)$$

For spin ½ particles (such as electrons) s = ½, the number of states in the energy range dE at E is

$$g(E)dE = \frac{4\pi V}{h^3}(2m)^{3/2} E^{1/2} dE \quad (2.1.7)$$

The function $g(E)$ is called the **density of states** and is defined as the number of quantum states per unit energy range at energy E and is given by

$$g(E) = (2s+1)\frac{2\pi V}{h^3}(2m)^{3/2} E^{1/2} \quad (2.1.8)$$

In $6N$-dimensional γ-space of a N-particle system, the hypervolume h^{3N} represents a quantum state of the system. The number of cells (states) in a volume element $dq_1 \ldots dq_{3N}$, $dp_1 \ldots\ldots\ldots dp_{3N}$ at point (q, p) in γ-space is $\frac{dq_1 \ldots\ldots dq_{3N} . dp_1 \ldots\ldots dp_{3N}}{h^{3N}}$. The total number of accessible states in γ-space is

$$g(q,p)d^{3N}qd^{3N}p = \frac{1}{h^{3N}} \int \ldots\ldots\ldots \int dq_1 \ldots\ldots dq_{3N} . dp_1 \ldots\ldots\ldots dp_{3N}$$

$$= \frac{1}{h^{3N}} \int dq_1 \ldots\ldots\ldots dq_{3N} . \int dp_1 \ldots\ldots\ldots dp_{3N}$$

$$= \frac{1}{h^{3N}} . V^N . \int dp_1 \ldots\ldots dp_{3N}$$

$$= \frac{V^N}{h^{3N}} \int d^3 p_1 . d^3 p_2 \ldots\ldots\ldots d^3 p_N$$

$$= \frac{V^N}{h^{3N}} \int d^{3N} p \quad (2.1.9)$$

3

Ensemble Formulation of Statistical Mechanics

3.1 ENSEMBLE

The method of ensemble in statistical physics was introduced in 1902 by the American physicist J.W.Gibbs. Consider a system consisting of N molecules with total energy E enclosed in a vessel of volume V. The macroscopic state of the system is described by pressure P, volume V and energy E. With passage of time the coordinates and momenta of molecules and hence the microscopic states of the system undergo continuous change. Meaning there by, a macrostate is realized through an enormously large number of microstates. Now imagine a large number (possibly infinite) of systems, which are exactly identical in structure to the system of interest, but suitably randomized in microscopic states such that they represent at one time the possible states of the actual system attained in the course of time. This mental collection of similar non-interacting, independent systems is called an *ensemble*. All the members of an ensemble, which are identical in feature like N, V and E are called the elements or systems. These elements, although identical in structure are randomized in the sense that they differ from one another in the coordinates and momenta of the individual molecules *i.e.*, the elements differ in their unobservable microscopic states.

The state of an element (system) of the ensemble can be specified by the $3N$ canonical coordinates $q_1, q_2, ...q_{3N}$, and $3N$ canonical momenta $p_1, p_2, ...p_{3N}$ of the N molecules. The $6N$ dimensional space spanned by the $3N$ spatial and $3N$ momentum coordinates is called γ-space of the system. An element of the ensemble is represented by a point and the ensemble is represented by a distribution of points in γ-space usually a continuous distribution.

Instead of denoting the spatial and momentum coordinates of an N-particle system by $q_1,....q_{3N}$, and $p_1,....p_{3N}$, it is convenient to denote them by $q_1, ...q_f, p_1,......p_f$ respectively. Of course $f = 3N$. The system is said to have f degrees of freedom. In $2f$ (= $6N$) dimensional γ-space having $2f$ rectangular axes, one for each of spatial coordinates $q_1, ...q_f$ and one for each of corresponding momentum coordinates $p_1,....p_f$, the state of the system is represented by a point. In this space the ensemble of system looks like cloud of points. The ensemble may be conveniently described by a density function $\rho(p, q, t)$ where (p, q) is an abbreviation for $p_1, p_2, ... p_f, q_1, q_2, ...q_f$, so defined that $\rho(p, q, t)\, d^fp\, d^fq$ is the number of representative points which at time t are contained in the infinitesimal volume element $d^fp\, d^fq$ of γ-space centered about the point (p, q). An ensemble is completely specified by $\rho(p, q, t)$. It is to be emphasized that the members of an ensemble are the mental copies of a system and do not interact with one another.

Each element of the ensemble is a quantum mechanical system of N interacting molecules in a container of volume V. The value of N and V along with the force law between the molecules, are sufficient to determine the energy eigenvalues and the quantum states of the Schrodinger equation along with their associated degeneracies. These energies are the only energies available to the system. The energy E of the system must be one of these energy eigenvalues. We shall further restrict our ensemble to obey the principle of *equal a priori probabilities.* That is to say we require each and every quantum state is represented an equal number of times in the ensemble. Since we have no information to consider any one of the quantum states to be more important than any other, we must treat each of them equally *i.e.*, we must utilize the principle of *equal a priori probability.*

3.2 DENSITY OF DISTRIBUTION IN γ-SPACE

The condition of an ensemble can be described in terms of the density ρ with which the representative points are distributed in phase space. For an ensemble consisting of very large number of systems, the distribution of representative points is continuous and the density of phase points in γ-space can be treated as a continuous function. With passage of time, the microstates of the systems undergo change and their representative points move in γ-space from one region to another. So the density ρ of phase points is function of $q_1,....q_f,\ p_1,......p_f$ and time t and may be written as

$$\rho = \rho\ (q_1,.....q_f,\ p_1,......p_f,\ t) = \rho\ (q,\ p,\ t) \tag{3.2.1}$$

The meaning of ρ is such that

$$\rho\ dq_1....dq_f\ dp_1......dp_f \text{ or } \rho\ d\Gamma \tag{3.2.2}$$

represents the number of systems in infinitesimal hypervolume $d\Gamma = dq\ dp = dq_1...dq_f\ .dp_1.....dp_f$ located at point $(q,\ p)$. The number of systems dM in hypervolume $d\Gamma$ is

$$dM = \rho\ d\Gamma \tag{3.2.3}$$

and the total number of systems in the ensemble is

$$M = \int \rho dT \tag{3.2.4}$$

Where the integration is over the whole phase space. The average value of a physical quantity Q $(q,\ p)$ is given by

$$(Q) = \frac{\int Q(q,p)\rho(q,p,t)dT}{\int \rho(q,p,t)dT} \tag{3.2.5}$$

3.3 PRINCIPLE OF EQUAL A PRIORI PROBABILITY

To specify the microscopic state of a system, the phase space of the system is subdivided into small cells of equal size. Each cell represents a microscopic state of the system. According to the postulate of equal a priori probability, *an isolated system in equilibrium is equally likely to be in any of its accessible microscopic states satisfying the macroscopic conditions of the system.*

A many particle system, according to quantum mechanics, possesses discrete energy levels and discrete quantum states. It is found that many distinct quantum states correspond to the same energy levels. The number of different quantum states having the same energy is called the degeneracy of that energy level. The particles constituting the system are distributed among the various energy states. The specification of the macroscopic parameters, such as total energy E, volume V, the total number N of particles of the system, defines a particular macroscopic state of the system. Let the allowed energy levels be denoted by $\varepsilon_1,\ \varepsilon_2,\ ....\ \varepsilon_i$ and the occupation number of these energy levels by $n_1,\ n_2,\ .....n_i$. Let the system obey the constraints

$$\sum_i n_i = N \qquad and \qquad \sum_i n_i \varepsilon_i = E \tag{3.3.1}$$

There can be a large number of different ways in which the total energy E of the system can be distributed among N particles constituting the system. Each of these different ways specifies a particular microscopic state of the given system. To a given macroscopic state there may be a large number of microscopic states. According to the principle of *equal a priori probability, when a system is in statistical equilibrium, all the microstates are equally probable.*

There is no direct proof of this postulate. It does not contradict any known laws of mechanics. All calculations based on this postulate have yielded results that are in very good agreement with observations. The validity of this postulate can therefore be accepted with great confidence as the basis of our theory.

3.4 ERGODIC HYPOTHESIS

In statistical mechanics we often deal with average or the mean of a quantity. The average of a physical quantity can be determined in two ways:

(*i*) One could consider an ensemble of a large number of identical systems and average the physical quantity over all these systems at one instant of time to determine its ensemble average.

(*ii*) A system could be followed over a very long period of time, during which the physical quantity of the system takes different values. The average of the physical quantity over the long period gives the time averaged value of the quantity.

According to ergodic hypothesis the mean over the ensemble is equal to the mean over time. So far, there is no proof of the validity of this statement in the general case and is taken as one of the basic assumptions of statistical physics. The *ergodic hypothesis* and the *principle of equal a priori probability* are the main postulates that are employed for studying the properties of an ensemble.

3.5 LIOUVILLE'S THEOREM

Consider an isolated system specified by spatial and momentum coordinates $q_1, \ldots\ldots q_f, p_1, \ldots\ldots p_f$. In $2f$ dimensional γ-space the ensemble of systems appears as a cloud of points. In course of time the phase points move in γ-space because of change in position and momentum coordinates. This will result in change in the distribution density $\rho\ (q, p, t)$ of phase points. Let us define $\rho\ (q, p, t)$ such that $\rho\ (q, p, t)\ dq_1 \ldots dq_f dp_1 \ldots\ldots dp_f$ represents the number of phase points (systems) in hypervolume $d\Gamma = dqdp$ located at point (q, p). The Liouville's theorem gives the rate of change of density $\rho\ (q, p, t)$ at a fixed point in γ-space.

Consider an element of hypervolume of phase space located between q_1 and $q_1 + dq_1$, q_2 and $q_2 + dq_2 \ldots\ldots q_f$ and $q_f + dq_f$ and p_1 and $p_1 + dp_1$, p_2 and $p_2 + dp_2 \ldots\ldots\ldots p_f$ and $p_f + dp_f$. The volume of this element is $d\Gamma = dq_1 \ldots\ldots dq_f \,.\, dp_1 \ldots\ldots dp_f$. The coordinates and momenta of the phase points vary according to the Hamilton's equations of motion

$$\dot{q}_i = \frac{\partial H}{\partial p_i}, \qquad \dot{p}_i = -\frac{\partial H}{\partial q_i} \tag{3.5.1}$$

where $H = H\ (q_1, \ldots q_f, p_1, \ldots\ldots p_f)$ is Hamiltonian of the system. The change in q's and p's results a change in the number of phase points in the element of hypervolume. In time dt, the change in the number of phase points within this hypervolume of phase space is

$$\left(\frac{\partial \rho}{\partial t} dt\right) d\Gamma$$

This change is equal to the difference in the number of phase points entering and leaving this volume in time dt. The number of phase points entering this volume in time dt through the face located at q_1 = constant is

$$\rho(q, p, t)(\dot{q}_1 dt)\, dq_2 \ldots\ldots\ldots\ldots dq_f\, dp_1 \ldots\ldots\ldots\ldots dp_f$$

The number of phase points leaving through the opposite face located at $q_1 + dq_1$ = constant is

$$\left\{\rho\dot{q}_1 + \frac{\partial}{\partial q_1}(\rho\dot{q}_1)\, dq_1\right\} dt\, dq_2 \ldots\ldots\ldots dq_f\, dp_1 \ldots\ldots\ldots dp_f$$

The net number of phase points entering the hypervolume element in time *dt*

$$= \rho\dot{q}_1 dt dq_2 \ldots dq_f dp_1 \ldots\ldots dp_f - \left\{\rho\dot{q}_1 + \frac{\partial(\rho\dot{q}_1)}{\partial q_1} dq_1\right\} dt dq_2 \ldots\ldots dq_f dp_1 \ldots\ldots dp_f$$

$$= -\frac{\partial}{\partial q_1}(\rho\dot{q}_1)\, dt dq_1 \ldots\ldots dq_f dp_1 \ldots\ldots\ldots dp_f$$

$$= -\frac{\partial(\rho\dot{q}_1)}{\partial q_1} dt\, d\Gamma$$

The total net increase in time dt of the number of phase points in the hypervolume of phase space is obtained by summing the net number of phase points entering the hypervolume through all the faces labeled by $q_1 \ldots\ldots q_f p_1 \ldots\ldots\ldots p_f$. Thus one obtains

$$\frac{\partial \rho}{\partial t} dt\, d\Gamma = -\left[\sum_{i=1}^{f} \frac{\partial}{\partial q_i}(\rho\dot{q}_i) + \sum_{i=1}^{f} \frac{\partial}{\partial p_i}(\rho\dot{p}_i)\right] dt\, d\Gamma$$

or

$$\frac{\partial \rho}{\partial t} = -\sum_{i=1}^{f}\left[\frac{\partial}{\partial q_i}(\rho\dot{q}_i) + \frac{\partial}{\partial p_i}(\rho\dot{p}_i)\right] \qquad (3.5.2)$$

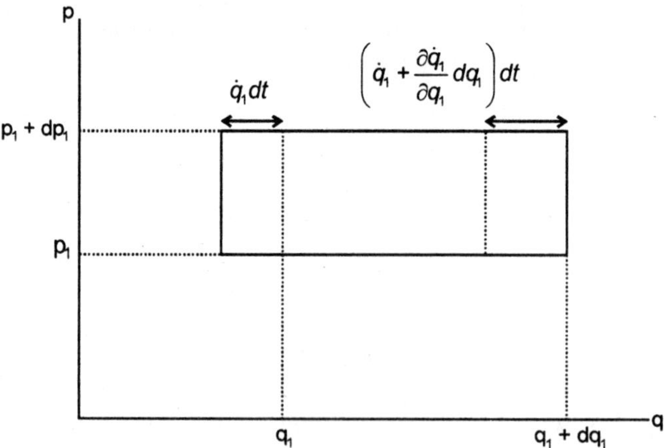

Fig. 3.5 A volume element in phase space

Eq. (3.5.2) can be written as

$$\frac{\partial \rho}{\partial t} = -\sum_{i=1}^{f}\left[\left(\frac{\partial \rho}{\partial q_i}\dot{q}_i + \frac{\partial \rho}{\partial p_i}\dot{p}_i\right) + \rho\left(\frac{\partial \dot{q}_i}{\partial q_i} + \frac{\partial \dot{p}_i}{\partial p_i}\right)\right] \tag{3.5.3}$$

Making use of Hamilton's equations we have

$$\frac{\partial \dot{q}_i}{\partial q_i} + \frac{\partial \dot{p}_i}{\partial p_i} = \frac{\partial^2 H}{\partial q_i \partial p_i} - \frac{\partial^2 H}{\partial p_i \partial q_i} = 0 \tag{3.5.4}$$

In view of (3.5.4) Eq.(3.5.3) reduces to

$$\left(\frac{\partial \rho}{\partial t}\right)_{q,p} = -\sum_{i=1}^{f}\left(\frac{\partial \rho}{\partial q_i}\dot{q}_i + \frac{\partial \rho}{\partial p_i}\dot{p}_i\right) \tag{3.5.5}$$

Making use of Hamilton's equations of motion we can write (3.5.5) as

$$\left(\frac{\partial \rho}{\partial t}\right)_{q,p} = -\sum_{i=1}^{f}\left(\frac{\partial \rho}{\partial q_i}\frac{\partial H}{\partial p_i} - \frac{\partial \rho}{\partial p_i}\frac{\partial H}{\partial q_i}\right) \tag{3.5.6}$$

Equation (3.5.5) or (3.5.6) is known as the Liouville theorem. It gives the rate of change of density at a fixed point in γ-space. In view of the following results

$$\sum_{i}^{f}\frac{\partial \rho}{\partial q_i}\dot{q}_i = \sum_{i}^{f}\frac{\partial \rho}{\partial q_i}\frac{\partial q_i}{\partial t} = \left(\frac{\partial \rho}{\partial t}\right)_{p,t}$$

and

$$\sum_{i}^{f}\frac{\partial \rho}{\partial p_i}\dot{p}_i = \sum_{i}^{f}\frac{\partial \rho}{\partial p_i}\frac{\partial p_i}{\partial t} = \left(\frac{\partial \rho}{\partial t}\right)_{q,t}$$

we can write Eq. (3.5.5) as follows

$$\left(\frac{\partial \rho}{\partial t}\right)_{q,p} + \sum_{i=1}^{f}\left(\frac{\partial \rho}{\partial q_i}\dot{q}_i + \frac{\partial \rho}{\partial p_i}\dot{p}_i\right) = 0$$

$$\left(\frac{\partial \rho}{\partial t}\right)_{q,p} + \left(\frac{\partial \rho}{\partial t}\right)_{p,t} + \left(\frac{\partial \rho}{\partial t}\right)_{q,t} = 0 \tag{3.5.6}$$

$$\frac{d\rho}{dt} = 0 \tag{3.5.7}$$

Thus the total derivative of density ρ (q, p, t), which is a measure of the rate of change of ρ in the immediate vicinity of a *moving* phase point (q and p changing) in γ-space, is zero. In other words the density of a group of phase points remains constant along their trajectories in the γ-space. The distribution of phase points moves in γ-space like an incompressible fluid. Gibbs called this conclusion the *principle of the conservation of density in phase.*

From Eq. (3.5.7) we can obtain another fundamental principle of statistical mechanics. Consider a region in γ-space which, although finite, is small enough for the density ρ to be treated as uniform throughout; if the hypervolume of the region is $\delta\Gamma$, the number δM of the phase points in this region will be given by

$$\delta M = \rho\, \delta\Gamma \tag{3.5.8}$$

On differentiating this expression with respect to time t, it is seen that

$$\frac{d}{dt}(\delta M) = \frac{d\rho}{dt}\delta\Gamma + \rho\frac{d(\delta\Gamma)}{dt} \tag{3.5.9}$$

If it is supposed that the boundaries of the region under consideration are permanently determined by the phase points that were originally on the surface, then no phase points can enter or leave this region. In other words, the points on the outer surface act like a continuous thin skin by which all the points in the region are enclosed. The hypersurface enclosing the region changes its shape and moves about on gamma space due to the flow of phase points. Further, since each phase point represents a definite system, these points can neither be created nor destroyed. So $\frac{d}{dt}(\delta M) = 0$. Eq. (3.5.9) then becomes

$$\frac{d\rho}{dt}\delta\Gamma + \rho\frac{(\delta\Gamma)}{dt} = 0 \tag{3.5.10}$$

Since $\frac{d\rho}{dt} = 0$, we have

$$\frac{d(\delta\Gamma)}{dt} = 0 \tag{3.5.11}$$

This means that the volume or *extension-in-phase* in γ-space of the particular region, occupied by a definite number of phase points, does not change with time. Since every finite arbitrary extension-in-phase may be regarded as composed of infinitesimal parts, the result may be generalized. This theorem, mathematically expressed by equation (3.5.11) is called the *principle of conservation of extension in phase.*

3.6 STATISTICAL EQUILIBRIUM

An ensemble is said to be in statistical equilibrium if the density of phase points is independent of time at all points in γ-space *i.e.*,

$$\left(\frac{\partial\rho}{\partial t}\right)_{q,p} = 0 \quad \text{for all } q\text{'s and } p\text{'s.}$$

Consider an ensemble of conservative systems for which energy E is constant in time and is function of q's and p's. Thus

$$\rho = \rho\,(E), \quad E = E\,(q,p) \text{ and } \frac{dE}{dt} = 0 \tag{3.6.1}$$

Therefore

$$\frac{\partial\rho}{\partial p_i} = \frac{d\rho}{dE}\cdot\frac{\partial E}{\partial q_i}, \qquad \frac{\partial\rho}{\partial p_i} = \frac{d\rho}{dE}\cdot\frac{\partial E}{\partial p_i} \tag{3.6.2}$$

According to Liouville's theorem

$$\left(\frac{\partial \rho}{\partial t}\right)_{q,p} = -\sum_{i}^{f}\left(\frac{\partial \rho}{\partial q_i}\dot{q}_i + \frac{\partial \rho}{\partial p_i}\dot{p}_i\right) \tag{3.6.3}$$

Making use of (3.6.2) in (3.6.3) we have

$$\left(\frac{\partial \rho}{\partial t}\right)_{q,p} = -\frac{d\rho}{dE}\sum_{i}^{f}\left(\frac{\partial E}{\partial q_i}\dot{q}_i + \frac{\partial E}{\partial p_i}\dot{p}_i\right) \tag{3.6.4}$$

Since $E = E\ (q,\ p)$ and $dE/dt = 0$ we have

$$\frac{dE}{dt} = \sum_{i}^{f}\left(\frac{\partial E}{\partial q_i}\dot{q}_i + \frac{\partial E}{\partial p_i}\dot{p}_i\right) = 0 \tag{3.6.5}$$

From (3.6.4) and (3.6.5)

$$\left(\frac{\partial \rho}{\partial t}\right)_{q,p} = 0 \text{ for all q's and p's.} \tag{3.6.6}$$

Thus an ensemble is in statistical equilibrium if density of phase points (or the probability of finding the phase points) in the various regions of γ-space is independent of time. This means that every portion of the phase space continues to contain the same number of phase points at all times. Under these conditions, the average values of the properties of the systems in the ensemble also do not change with time.

3.7 ENSEMBLE FORMULATION OF STATISTICAL MECHANICS

There are three kinds of formulations of statistical physics. They are:

(*i*) Microcanical Ensemble

(*ii*) Canonical Ensemble

(*iii*) Grand Canonical Ensemble.

3.7.1 Microcanonical Ensemble

A collection of a very large number (possibly infinite) of closed *isolated identical systems* is called a microcanonical ensemble. The elements of the microcanonical ensemble are isolated in the sense that its energy is constant of motion. This is clearly an idealization, for we never deal with truly isolated systems in the laboratory. The walls of the container containing the system will be perfectly insulating. In $6N$ dimensional γ-space of the system, each point represents a state of the system and vice versa. The locus of all points in γ-space satisfying the condition E = constant defines a surface called the ergodic surface of energy E. As the state of the system evolves in time according to Hamilton's equations of motion, the representative point traces out a path in γ-space. This path always stays on the same energy surface because by definition energy is conserved. We cannot specify exactly the energy of a system. However, we can certainly specify the energy within some narrow range. We can then select two neighboring ergodic surfaces, one at E and the other at $E + \Delta E$. The phase point of a conservative system remains always on the same ergodic surface. In $6N$ dimensional γ-space the microcanonical ensemble, whose each member has energy between

E and $E + \Delta E$, is represented by points that lie between two ergodic surfaces of energies E and $E + \Delta E$. A microcanonical ensemble may be represented by distribution of points in γ-space characterized by a density function $\rho(p, q, t)$ defined in such a way that $\rho(p, q, t)\, d^{3N}p\, d^{3N}q$ gives the number of representative points contained in the volume element $d^{3N}p\, d^{3N}q$ located at point (p, q) in γ-space at the instant t. For microcanonical ensemble the density function satisfies the condition

$$\begin{aligned} \rho(p, q) &= 1 \text{ if } E < H(p, q) < E + \Delta E \\ &= 0 \text{ otherwise} \end{aligned} \tag{3.7.1}$$

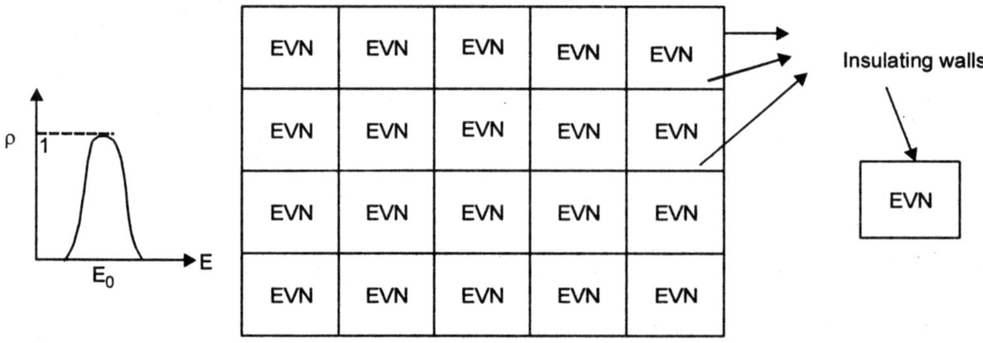

Fig. 3.7.1 Microcanonical ensemble (*EVN* fixed)

The average of a physical quantity $R(p,q)$ is defined by

$$\langle R \rangle = \frac{\int d^{3N}p\, d^{3N}q\ R(p,q)\, \rho(p,q)}{\int d^{3N}p\, d^{3N}q\ \rho(p,q)} \tag{3.7.2}$$

The statistical description of a system in this approach is given in terms of statistical weight Ω (*EVN*) which leads to the thermodynamic description in terms of entropy S through Boltzmann equation $S = k \ln \Omega\ (EVN)$.

3.7.2 Canonical Ensemble and Canonical Distribution

Consider a system, with fixed number of particles N and volume V, immersed in a heat bath at temperature T. When the equilibrium is reached the temperature of the system attains the value T. The energy of the system assumes different values due to exchange of heat with the heat bath. An ensemble of such systems can mentally be constructed as follows. A system is enclosed in a container of volume V, the walls of the container are heat conducting but impermeable to the passage of particles. A very large number, say M, of such identical systems are, then, supposed to be immersed in a very large heat bath at temperature T. When the equilibrium is reached, the entire ensemble is at the same temperature T. Each system of the ensemble has the same values of N, V, and T. The aggregate of such systems constitutes canonical ensemble. Now, the entire canonical ensemble is enclosed within a non-conducting wall. All the systems of the canonical ensemble are at the same temperature but the different systems have different energies.

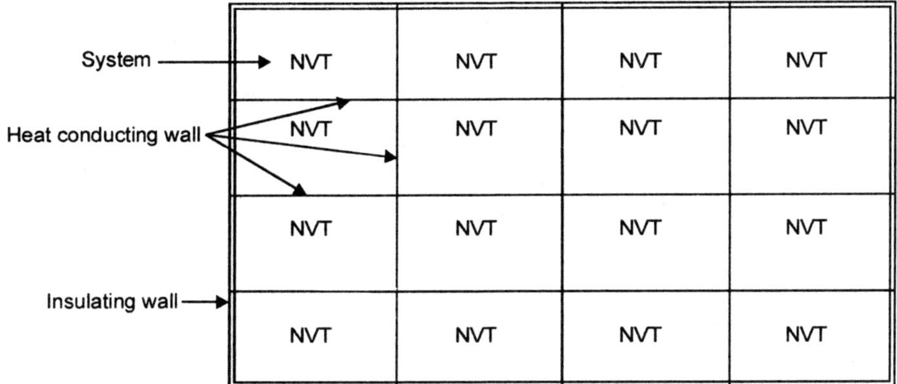

Fig. 3.7.2 Canonical ensemble

In a system consisting of gas molecules, the individual molecules may be treated as a system and the rest of the molecules as a heat bath. Energy of different molecules (systems) have different values.

Gibbs Canonical Distribution

The Gibbs canonical distribution gives the probabilities of occurrence of different energy states of systems constituting the canonical ensemble. To obtain an expression for this probability consider a system A placed in a heat bath, which we denote by A'. Our system of interest A and the heat bath A' constitute a composite system A*. The composite system is totally isolated. The system A and A' are free to exchange energy but the temperature of the heat bath remains constant because of its large size. In thermal equilibrium the temperature of A is the same as that of the heat bath. The energy of A is not fixed. The macroscopic state of the system is specified by N, V, T.

We assume that the sytem A possesses a discrete set of microstates labeled by 1, 2, 3, ...r....and in these states the system has energy ε_1 ε_2, ε_3,........ ε_r....... It is possible that many distinct microstates (quantum states) have the same energy. We also assume that all energy levels are discrete and are separated by small interval δE. The magnitude of δE is supposed to be large enough to contain many energy levels. Let E, E' and E^* denote the energies of the systems A, A' and A^*. The system A has energy E means that its energy lies anywhere between E and E + δE. Similar statements apply to energies of A' and A*. Since A*is enclosed within a heat insulating walls, we have

$$E + E' = E^* = \text{constant}$$

or

$$E' = E^* - E \tag{3.7.3}$$

Let $\Omega'(E) = \Omega'(E^* - E)$ denote the number of states accessible to A' when its energy is equal to E' (= E^* – E). When the system A is in its state r with energy ε_r, A' must have energy $\varepsilon' = \varepsilon^* - \varepsilon_r$. When A is in this one definite state r, the number of states accessible to the isolated system A^* is equal the number of states $\Omega'(\varepsilon^* - \varepsilon_r)$, which are accessible to A'. According to the fundamental postulate of statistical mechanics, all the accessible states of an isolated system are equally probable. Therefore, the probability that the system A is in the state r is proportional to the corresponding number of states accessible to the system A^*.

$$p_r \propto \Omega'(\varepsilon') = \Omega'(\varepsilon^* - \varepsilon_r) \tag{3.7.4}$$

$\Omega'(\varepsilon')$ is a rapidly varying function. It is more convenient to work with more slowly varying function $\ln\Omega'(\varepsilon')$ or $\ln\Omega'(\varepsilon^* - \varepsilon_r)$. Expanding $\ln\Omega'(\varepsilon^* - \varepsilon_r)$ about the value ε^* we have

$$\ln \Omega' (\varepsilon^* - \varepsilon_r) = \ln \Omega'(\varepsilon^*) - \left(\frac{\partial \ln \Omega'}{\partial \varepsilon'}\right)\varepsilon_r - \text{higher order terms}$$

$$= \ln \Omega' (\varepsilon^*) - \beta\varepsilon_r \tag{3.7.5}$$

where
$$\beta = \frac{\partial \ln \Omega'}{\partial \varepsilon'} \tag{3.7.6}$$

The derivative is evaluated at fixed energy $\varepsilon' = \varepsilon^*$. Eq. (3.7.5) can be written as

$$\ln\Omega'(\varepsilon^* - \varepsilon_r) - \ln\Omega'(\varepsilon^*) = -\beta\,\varepsilon_r$$

or
$$\Omega'(\varepsilon^* - \varepsilon_r) = \Omega'(\varepsilon^*)\exp(-\beta\,\varepsilon_r) \tag{3.7.7}$$

$\Omega'(\varepsilon^*)$ is constant and is independent of r. Making use of (3.7.7) in (3.7.4) we get

$$p_r = C\exp(-\beta\varepsilon_r) \tag{3.7.8}$$

where C is a proportionality constant and is independent of r. The constant C can be determined by the normalization condition that the system must have probability unity of being in some one of its states *i.e.*,

$$\sum_r p_r = 1$$

where the summation is over all states of A irrespective of energy. That is

$$C\sum_r \exp(-\beta\varepsilon_r) = 1$$

This gives
$$C = \frac{1}{\sum_r \exp(-\beta\varepsilon_r)}$$

Eq. (3.7.8) then becomes

$$p_r = \frac{1}{\sum_r \exp(-\beta\varepsilon_r)}\exp(-\beta\varepsilon_r) = \frac{1}{Z}\exp(-\beta\varepsilon_r) \tag{3.7.9}$$

where
$$Z = \sum_r \exp(-\beta\varepsilon_r) \tag{3.7.10}$$

Eq. (3.7.9) is a very general result of fundamental importance in statistical mechanics. The exponential factor $\exp(-\beta\varepsilon_r)$ is called *Boltzmann factor* and the corresponding probability distribution is known as the *canonical distribution*.

The quantity Z defined by $Z = \sum_r \exp(-\beta\varepsilon_r)$ is called the *partition function.*

The quantity β can be shown equal to $1/kT$, k is Boltzmann constant and T is absolute temperature of the system.

It may happen that different states may have the same energy. Let $g(\varepsilon_r)$ be the number of distinct quantum states having the energy ε_r. Then the expression for the partition function becomes

$$Z = \sum_{diff.energy\ levels} g(\varepsilon_r) e^{-\beta\varepsilon_r} \tag{3.7.11}$$

The probability $p(\varepsilon_r)$ that the system be in a state with energy ε_r is then given by

$$p(\varepsilon_r) = \frac{1}{Z} g(\varepsilon_r) \exp(-\beta\varepsilon_r) \tag{3.7.12}$$

The expression in Eq.(3.7.12) may be extended to a system composed of molecules, which are distinguishable. Suppose that the gas is in equilibrium at temperature T. If the gas is dilute enough so that the molecules are distinguishable, we can focus attention on a particular molecule of the gas and regard it as a small system in thermal contact with a heat reservoir consisting of all the remaining molecules of the gas. The probability of finding the molecule in any one of its quantum state r where its energy is ε_r is then given by the canonical distribution

$$p(\varepsilon_r) = \frac{g(\varepsilon_r) e^{-\beta\varepsilon_r}}{\sum_r g(\varepsilon_r) e^{-\beta\varepsilon_r}} = \frac{1}{Z} . g(\varepsilon_r) e^{-\beta\varepsilon_r} \tag{3.7.13}$$

Another interpretation of probability $p(\varepsilon_r)$ is that if $\langle n_r \rangle$ is the mean number of molecules occupying the energy level ε_r then

$$p(\varepsilon_r) = \frac{\langle n_r \rangle}{N}, \quad \text{N = total number of molecules in the gas}$$

Therefore
$$p(\varepsilon_r) = \frac{\langle n_r \rangle}{N} = \frac{1}{Z} g(\varepsilon_r) . e^{-\beta\varepsilon_r} \tag{3.7.14}$$

whence
$$\langle n_r \rangle = \frac{N}{Z} g(\varepsilon_r) e^{-\beta\varepsilon_r} \tag{3.7.15}$$

A system in canonical ensemble in equilibrium with a reservoir has access to all its possible states, and the probability varies exponentially with the energy of the state. The Boltzmann probability is a probability per quantum state. To get probability per unit range of energy, $f(\varepsilon)$ we must multiply p_r by the density of states.

$$f(\varepsilon) d\varepsilon = p(\varepsilon) . g(\varepsilon) d\varepsilon \tag{3.7.16}$$

It is true that $p(\varepsilon)$ falls rapidly as energy rises but initially, $g(\varepsilon)$ rises even more rapidly, so that the energy distribution has a peak far above the ground state energy , even though the individual states near the peak have extremely low probabilities.

Average Energy of Particle

If a molecule is found with probability p (ε_r) in a state r of energy ε_r, then its mean energy is given by

$$\langle \varepsilon \rangle = \sum_r p_r \varepsilon_r = \frac{\sum_r \varepsilon_r \; e^{-\beta \varepsilon_r}}{\sum_r e^{-\beta \varepsilon_r}} \tag{3.7.17}$$

Now $$\sum_r \varepsilon_r \; e^{-\beta \varepsilon_r} = -\sum_r \frac{\partial}{\partial \beta} e^{-\beta \varepsilon_r} = -\frac{\partial}{\partial \beta}\left(\sum_r e^{-\beta \varepsilon_r}\right) = -\frac{\partial Z}{\partial \beta}$$

Therefore $$\langle \varepsilon \rangle = -\frac{1}{Z}\frac{\partial Z}{\partial \beta} = -\frac{\partial \ln Z}{\partial \beta} \tag{3.7.18}$$

Hence total energy of the N-particle system is

$$E = N\langle \varepsilon \rangle = -N\frac{\partial \ln Z}{\partial \beta} = NkT^2 \frac{\partial \ln Z}{\partial T} \tag{3.7.19}$$

3.8 THE EQUIPARTITION THEOREM

Consider a system whose state is described classically in terms of f coordinates $q_1, \ldots q_f$ and f corresponding momenta $p_1, \ldots p_f$. Let the energy of the system be function of these coordinates. For most of the systems, which we shall be dealing with, the energy can be written as

$$\text{E}\,(q, p) = \varepsilon_{\text{I}}\,(p_i) + \varepsilon'(q_1 \ldots q_f, p_1 \ldots p_f) \tag{3.8.1}$$

Where the first term is function of the particular momentum p_i only and the second term may depend on all coordinates and momenta except p_i. [For example, the energy of a harmonic oscillator can be written as

$$E = \varepsilon_{kinetic} + \varepsilon_{potential} = \frac{p^2}{2m} + \frac{1}{2}\alpha q^2$$

The first term depends on momentum p only and the second term on coordinate q, α is force constant.] Suppose that the system under consideration is in thermal equilibrium with a heat reservoir at temperature T. The probability of finding the system with its coordinates lying in the range q and $q + dq$ and momenta in the range p and $p + dp$ is given by canonical distribution

$$P(q, p) = \frac{e^{-\beta E(q,p)}}{\int e^{-\beta E(q,p)} dqdp} \tag{3.8.2}$$

The mean value of energy ε_{I} (p_i) is given by

$$\langle \varepsilon_i \rangle = \frac{\int e^{-\beta E(q,p)} \varepsilon_i(p_i) dqdp}{\int e^{-\beta E(q,p)} dqdp} \tag{3.8.3}$$

where the integrals extend over all possible values of all coordinates [q] and momenta [p]. Substituting the expression for E from Eq. (3.8.1)) in (3.8.3) we have

$$\langle \varepsilon_i \rangle = \frac{\int e^{-\beta(\varepsilon_i+\varepsilon')}\ \varepsilon_i\ dqdp}{\int e^{-\beta(\varepsilon_i+\varepsilon')}\ dqdp}$$

$$= \frac{\int \varepsilon_i\ e^{-\beta\,\varepsilon_i} dp_i\ \int e^{-\beta\,\varepsilon'} dqdp}{\int e^{-\beta\,\varepsilon_i} dp_i\ \int e^{-\beta\,\varepsilon'} dqdp}$$

The primes on the last integrals indicate that these integrals extend over all the coordinates [q] and momenta [p] except p_i. The primed integrals in the numerator and denominator are equal and cancel out. Therefore

$$\langle \varepsilon_i \rangle = \frac{\int \varepsilon_i\ e^{-\beta\,\varepsilon_i}\ dp_i}{\int e^{-\beta\,\varepsilon_i}\ dp_i}$$

$$= \frac{-\dfrac{\partial}{\partial \beta}\left[\int e^{-\beta\,\varepsilon_i}\ dp_i\right]}{\int e^{-\beta\,\varepsilon_i}\ dp_i}$$

$$= -\frac{\partial}{\partial \beta}\left[\ln \int_{-\infty}^{\infty} e^{-\beta \varepsilon_i} dp_i\right] \qquad (3.8.4)$$

If ε_I is a quadratic function of p_i then

$\varepsilon_I = a\ p_i^2$ where a is a constant.

[For example the kinetic energy of a particle is quadratic function of momentum $\varepsilon = p^2/2m$.] Then the integral in Eq. (3.8.4) becomes

$$\int_{-\infty}^{\infty} e^{-\beta\,\varepsilon_i} dp_i = \int_{-\infty}^{\infty} e^{-a\beta\,p_i^2} dp_i = \frac{1}{\sqrt{\beta}} \int_{-\infty}^{\infty} e^{-ax^2} dx, \quad \text{where} \quad x = \sqrt{\beta}\,p_i$$

Hence $$\ln\left[\int_{-\infty}^{\infty} e^{-\beta\varepsilon_i} dp_i\right] = -\frac{1}{2}\ln\beta + \ln\left(\int_{-\infty}^{\infty} e^{-ax^2} dx\right)$$

The last integral does not involve β at all and its derivative with respect to β is zero. So we are left with

$$\langle \varepsilon_i \rangle = -\frac{\partial}{\partial \beta}\left\{-\frac{1}{2}\ \ln\ \beta\right\} = \frac{1}{2\beta} = \frac{1}{2}kT \qquad (3.85)$$

Thus we arrive at the conclusion *that if a system is in thermal equilibrium at temperature T, then each independent quadratic term in its energy contributes a mean value of energy equal to ½ kT to the total energy of the system.* This statement is known as the *theorem of equipartition of energy.*

4

Thermodynamic Functions

4.1 ENTROPY

Entropy is a very important thermodynamic function, which connects thermodynamics to statistical mechanics. It is known from thermodynamics that when a system, with constant volume and energy, is in equilibrium the entropy is maximum. On the other hand, according to statistical mechanics, such a system is in equilibrium when the total thermodynamic probability is a maximum. It appears, therefore, as suggested by Boltzmann, that there should be a relationship between entropy and thermodynamic probability. The thermodynamic probability Ω is defined as the number of microstates corresponding to the given macrostate. Let the entropy S and Ω be related through the expression

$$S = f(\Omega) \tag{4.1.1}$$

Consider two systems having entropies S_1 and S_2 and thermodynamic probabilities Ω_1 and Ω_2 respectively. In view of (4.1.1) we have

$$S_1 = f(\Omega_1) \text{ and } S_2 = f(\Omega_2).$$

Since the entropy is an additive quantity, the entropy of the combined system is equal to $S_{12} = S_1 + S_2$. The thermodynamic probability is a multiplicative quantity, therefore, the joint thermodynamic probability of the combined system is $\Omega_{12} = \Omega_1 \Omega_2$ and $S_{12} = f(\Omega_1 \Omega_2)$.

$$S_1 + S_2 = S_{12}$$

$$f(\Omega_1) + f(\Omega_2) = f(\Omega_1\Omega_2) \tag{4.1.2}$$

Differentiating (4.1.2) with respect to Ω_1 we have

$$f'(\Omega_1) = [f'(\Omega_1 \Omega_2)]\, \Omega_2 \tag{4.1.3}$$

Similarly, differentiating (4.1.2) with respect to Ω_2 we have

$$f'(\Omega_2) = [f'(\Omega_1 \Omega_2)]\, \Omega_1 \tag{4.1.4}$$

From (4.1.3) and (4.1.4)

$$\frac{f'(\Omega_1)}{f'(\Omega_2)} = \frac{\Omega_2}{\Omega_1}$$

$$\Omega_1 f'(\Omega_1)) = \Omega_2 f'(\Omega_2) = k, \text{ (= a constant, say)}$$

$$df(\Omega_1) = k\frac{d\Omega_1}{\Omega_1}$$

$$f(\Omega_1) = k \ln \Omega_1 + C_1$$

Similarly, we can have

$$f(\Omega_2) = k \ln \Omega_2 + C_2$$

C_1 and C_2 are constants. General form of these relations is

$$f(\Omega) = k \ln \Omega + C$$

or

$$S = k \ln \Omega + C \tag{4.1.5}$$

At absolute zero, any system is in most ordered state and this state has only one microstate *i.e.*, $\Omega = 1$ and this state is assigned zero entropy $S = 0$. The constant C in Eq. (4.1.5) comes out to be zero. So we have

$$S = k \ln \Omega \tag{4.1.6}$$

where k is Boltzmann constant.

4.2 ENTROPY IN TERMS OF PROBABILITY

The entropy is connected with the fact that in most thermodynamic states the system is not in a definite quantum state, but is spread over a large number of states according to some probability distribution.

To define entropy in terms of probability distribution consider an ensemble of very large number $M(M \to \infty)$ of systems. Let m_1 systems be in energy state ε_1, m_2 systems in energy state ε_2, and so on. The statistical weight Ω_M of the ensemble or the number of ways in which the systems are distributed among the various energy states is given by

$$\Omega_M\{m_i\} = \frac{M!}{m_1!m_2!.....} = \frac{M!}{\prod\limits_r m_r!} \tag{4.2.1}$$

The probability p_r that a system chosen at random will be in the state r with energy ε_r is

$$p_r = \frac{m_r}{M} \tag{4.2.2}$$

whence $m_r = M p_r$. The entropy of the ensemble is

$$S_M = k \ln \Omega_M = k\left[\ln \; M! - \sum_r m_r \ln m_r\right]$$

$$= k\left[\ln M! - \sum_r \ln m_r!\right] \approx k\left[M \ln M - \sum_r m_r \ln m_r\right]$$

$$= k\left[\sum_r M \ln M - \sum_r M p_r \ln m_r\right]$$

$$= kM\left[\ln M - \sum_r p_r \ln p_r M\right]$$

$$= kM\left[\ln M - \sum_r p_r (\ln p_r - \ln M)\right]$$

$$= -kM\sum_r p_r \ln p_r \tag{4.2.3}$$

The entropy is an extensive quantity and therefore entropy of a single system is

$$S = \frac{S_M}{M}$$

$$= -\ k\ \sum_r p_r \ln p_r \tag{4.2.4}$$

4.3 ENTROPY IN TERMS OF PARTITION FUNCTION

By definition $S = -k\sum_r p_r \ln p_r$

$$= -k\sum_r p_r(-\beta\varepsilon_r - \ln Z) \qquad \because\ p_r = \frac{e^{-\beta\varepsilon_r}}{Z}$$

$$= k\beta\left(\sum_r p_r\varepsilon_r\right) + k\ln Z\left(\sum_r p_r\right) \qquad \because \sum_r p_r = 1$$

$$= \frac{\bar{E}}{T} + k\ln Z \tag{4.2.5}$$

4.4 FREE ENERGY

In a mechanical system, such as a spring, the work done the system is stored in the system as potential energy and this energy may be recovered as work. In the similar way one can store energy in thermodynamic system, which can be recovered in the form of work. The energy, which can be stored and recovered, is called free energy. The four kinds of free energy that can be stored in thermodynamics system are: (*i*) Internal Energy E, (*ii*) Enthalpy $H = E + PV$, (*iii*) Helmholtz Free Energy F = E – TS and (*iv*) Gibb's Free Energy $G = E - TS + PV$.

The energy of a system also depends on the number of its constituent particles. When a particle leaves a system, it takes away a definite amount of energy with it. When it enters a system, it adds energy to it. To take into account the change in energy contributed by a particle we introduce a quantity, *chemical potential* μ, which is defined as the change in energy of the system associated with unit change in number of particles.

$$\mu = \left(\frac{\partial E}{\partial N}\right) \tag{4.4.1}$$

If a system is to be in equilibrium state, the temperature T, pressure P and chemical potential m must be the same throughout the system.

The law of conservation of energy for a system with variable number of particles can be written as

$$dE = TdS - PdV + \mu dN \tag{4.4.2}$$

where dN is the change in number of particles. For an isolated system at constant volume, which neither receives nor gives away heat, $TdS = 0$ and $dV = 0$. For such a system

$$dE = \mu dN$$

$$\mu = \left(\frac{\partial E}{\partial N}\right)_{S,V} \tag{4.4.3}$$

Hence the chemical potential represents the variation of the energy of an isolated system of constant volume brought about by a unit change in number of particles.

If E = constant and N = constant then (4.4.2) becomes

$$TdS = PdV$$

$$\left(\frac{\partial S}{\partial V}\right)_{N,E} = \frac{P}{T} \qquad \text{...(4.4.4)}$$

If V = constant, and E = constant, then (4.4.2) becomes

$$TdS = -\mu\, dN$$

$$\therefore \left(\frac{\partial S}{\partial N}\right)_{V,E} = -\frac{\mu}{T} \tag{4.4.5}$$

4.5 HELMHOLTZ FREE ENERGY IN TERMS OF PARTITION FUNCTION

Helmholtz free energy is defined by

$$F = E - TS \tag{4.5.1}$$

Substituting the expression for entropy $S = \dfrac{\bar{E}}{T} + k\ln Z$ in above equation we get

$$F = -kT\ln Z \tag{4.5.2}$$

From the definition of Helmholtz function $F = E - TS$, we have

$$dF = dE - TdS - SdT \tag{4.5.3}$$

The combined form of first and second law of thermodynamics is

$$TdS = dE + PdV \tag{4.5.4}$$

From (4.5.3) and (4.5.4) we have

$$dF = -SdT - PdV \tag{4.5.5}$$

whence

$$S = -\left(\frac{\partial F}{\partial T}\right)_{V,N} \tag{4.5.6}$$

and
$$P = -\left(\frac{\partial F}{\partial V}\right)_T \tag{4.5.7}$$

From (4.5.1) and (4.5.2) we have

$$S = k\,[\ln Z + \beta E] \tag{4.5.8}$$

4.6 THERMODYNAMIC FUNCTIONS IN TERMS OF PARTITION FUNCTION

Many thermodynamic properties of a system can be calculated from partition function Z. We give a list of few of them.

Total energy
$$E = -N\frac{\partial \ln Z}{\partial \beta} = NkT^2\frac{\partial \ln Z}{\partial T} \tag{4.6.1}$$

Entropy
$$S = k\ln\Omega = Nk\ln Z + \frac{E}{T} \tag{4.6.2}$$

Helmholtz Free Energy
$$F = E - TS = -NkT\ln Z \tag{4.6.3}$$

Gibb's Free Energy
$$G = E + PV - TS$$
$$= NkT^2\frac{\partial \ln Z}{\partial T} - NkT\left[\ln Z + \frac{1}{2}\right] \tag{4.6.4}$$

Pressure
$$P = -\left(\frac{\partial F}{\partial V}\right)_T = NkT\left(\frac{\partial \ln Z}{\partial V}\right)_T \tag{4.6.5}$$

Specific heat at constant volume

$$C_V = \left(\frac{\partial E}{\partial T}\right)_V = \frac{\partial}{\partial T}\left[NkT^2\frac{\partial \ln Z}{\partial T}\right] \tag{4.6.6}$$

5

Distribution Laws

5.1 MAXWELL–BOLTZMANN DISTRIBUTION

Consider a system of N identical *distinguishable* non-interacting particles confined to move freely in a vessel of volume V. The system is in equilibrium at temperature T. The total energy of the system is E. A monatomic ideal gas containing N molecules in a container of volume V is an example of our system. A macrostate of the system is specified by specifying how the total energy E of the system is distributed among N particles.

Let $\varepsilon_1, \varepsilon_2, \ldots\ldots\varepsilon_i \ldots$be the allowed single-particle energy levels and $g_1, g_2 \ldots\ldots g_i \ldots$the degeneracies of these energy levels respectively. The description of a macrostate is given by specifying the number of particles in each energy level ε_i. Let

n_1 be the number of particles in energy level ε_1 with degeneracy g_1

n_2 be the number of particles in energy level ε_2 with degeneracy g_2

....

n_i be the number of particles in energy level ε_i with degeneracy g_i and so on.

The set of numbers $\{n_1, n_2, \ldots.n_i\}$ defines a macrostate. Our objective is to determine $n_1, n_2, \ldots.n_i$. In classical (Maxwell-Boltzmann) statistics all the particles are assumed to be identical and *distinguishable* from one another. We wish to find the number of different arrangements or microstates corresponding to a macrostate. The required number of accessible microstates is equal to the number of ways, a given macrostate is realized by arranging the particles in different states. Now consider the *i-th* energy level, which is g_i fold degenerate. Each of the n_i particles can be placed in the g_i states in g_i different ways. The number of different ways of arranging n_i particles in g_i states is equal to the product

$$g_i . g_i . g_i . g_i . \ldots\ldots\ldots \; (n_i \text{ factors}) = g_i^{n_i}$$

This number $g_i^{n_i}$ gives number of microstates available to n_i particles, each having energy ε_i. The number of microstates available to the system such that n_1 particles are in energy level ε_1, n_2 are in energy level ε_2 ...and so on, is given by the product

$$g_1^{n_1} . g_2^{n_2} \ldots\ldots\ldots g_i^{n_i} = \prod_i g_i^{n_i}$$

Now we must take into account the possible permutations of the particles among the different energy levels. The number of permutations possible for N particles is $N!$. In other words, N particles can be arranged in $N!$ different sequences. Of these permutations some are irrelevant. When more than one particle is in an energy level, permuting them among themselves has no significance in this situation. Thus the n_i particles in the i-th level contribute $n_i!$ irrelevant permutations. If there are n_1 particles in level 1, n_2 particles in level 2 and so on, there are $n_1!\ n_2!\ n_3!$....irrelevant permutations. The number of ways N particles can be divided into groups of n_1, n_2, n_3 .... or the thermodynamic probability of a macrostate $\{n_1, n_2, ...n_i\}$ is

$$\frac{N!}{n_1!n_2!n_3!..........} \quad \text{or} \quad \frac{N!}{\prod_i n_i!}$$

The total number of distinct ways (accessible microstates) in which N particles can be distributed among possible energy levels such that n_1 particles are in level ε_1 with degeneracy g_1, n_2 are in level ε_2 with degeneracy g_2 and so on, is

$$\Omega_{MB} = \frac{N!}{n_1!n_2!n_3!......}(g_1)^{n_2}...... = N!\prod_i\left(\frac{g_i^{n_i}}{n_1!}\right) \tag{5.1.1}$$

According to the principle of equal a priori probabilities, all the microstates are equally probable. The most probable macro state specified by n_1, n_2 .....is one which corresponds to the maximum number of micro states. So to obtain most probable distribution we must maximize Ω_{MB} taking care of the conditions that the total number of particles and the total energy of the system is constant. These two restrictions are expressed as

$$n_1 + n_2 + n_3 +......... = \sum_i n_i = N \tag{5.1.2}$$

$$n_1\varepsilon_1 + n_2\varepsilon_2 + = \sum_i n_i\varepsilon_i = E \tag{5.1.3}$$

Mathematically it is more convenient to maximize $\ln \Omega_{MB}$ than Ω. So we first simplify $\ln \Omega_{MB}$. Now

$$\ln \Omega_{MB} = \ln N! + \sum_i (n_i \ln g_i - \ln n_i!)$$

Making use of Stirling formula $\ln n! = n \ln n - n$ we have

$$\ln \Omega_{MB} = N \ln N - N + \sum_i (n_i \ln g_i - n_i \ln n_i) + \sum_i n_i$$

$$= N \ln N + \sum_i n_i \ln g_i - \sum_i n_i \ln n_i \tag{5.1.4}$$

Since $\qquad \Sigma\, n_i = N.$

The most probable distribution is one for which Ω_{MB} or $\ln \Omega_{MB}$ is maximum. In other words a small change δn_i in any of the n_i's has no effect on the value of Ω_{MB}. We assume that n_i are continuous, so the condition of maximum $\ln \Omega_{MB}$ becomes

$$\frac{\partial \ln \Omega_{MB}}{\partial n_i} = 0 \quad \text{for each } n_i$$

If the change in $\ln \Omega_{MB}$ corresponding to change δn_i in n_i is $\delta \ln \Omega_{MB}$ then

$$\delta \ln \Omega_{MB} = \sum_i \ln g_i \, \delta n_i - \sum_i n_i \, \delta \ln n_i - \sum_i \ln n_i \, \delta n_i = 0$$

since N is constant. Now $\delta \ln n_i = \frac{1}{n_i} \delta n_i$. So

$$\delta \ln \Omega_{MB} = \sum_i \ln g_i \, \delta n_i - \sum_i \delta n_i - \sum_i \ln n_i \, \delta n_i = 0 \qquad (5.1.5)$$

Since the total number of particles is constant, $\sum_i \delta n_i = 0$. Therefore

$$\delta \ln \Omega_{MB} = \sum_i \ln g_i \, \delta n_i - \sum_i \ln n_i \, \delta n_i = 0 \qquad (5.1.6)$$

In order to incorporate the conditions of conservation of number of particles and of energy in above equation we use Lagrange's method of undetermined multipliers. From equations (5.1.2) and (5.1.3) we have

$$\sum_i \delta n_i = \delta n_1 + \delta n_2 + \dots\dots\dots\dots = 0 \qquad (5.1.7)$$

$$\sum_i \varepsilon_i \, \delta n_i = \varepsilon_1 \, \delta n_1 + \varepsilon_2 \, \delta n_2 \dots\dots\dots\dots = 0 \qquad (5.1.8)$$

Multiplying Eq.(5.1.7) by $-\alpha$ and Eq. (5.1.8) by $-\beta$, where α and β are quantities independent of n_i, and adding them to Eq.(5.1.6) we have

$$\sum_i (-\ln n_i + \ln g_i - \alpha - \beta \varepsilon_i) \delta n_i = 0 \qquad (5.1.9)$$

Eq. (5.1.9) will hold if the quantity in parentheses vanishes for each value of i. Hence

$$-\ln n_i + \ln g_i - \alpha - \beta \, \varepsilon_i = 0$$

$$n_i = g_i \, e^{-\alpha} \, e^{-\beta \varepsilon_i} \qquad (5.1.10)$$

The ratio n_i/g_i represents the average number of particles per state of the system and is called energy distribution function $f_{MB}(\varepsilon)$. *The distribution function represents the average number of particles in each of state of energy* ε *or the probability of occupancy of each of state of energy* ε.

$$f_{MB}(\varepsilon) = \frac{n_i}{g_i} = e^{-\alpha} \, e^{-\beta \varepsilon} \qquad (5.1.11)$$

In other words f_{MB} (ε) *gives the probability that a particle, selected randomly from the system, will have its energy* ε.

This is the classical (*M-B*) distribution function. It is applicable to system whose constituent particles are *distinguishable* and *don't obey Pauli's exclusion principle.* It is the most probable distribution of particles among the accessible energy levels at equilibrium for a system of constant total energy. For a system comprising of large number of particles the most probable distribution describes the actual behaviour of the system. In deriving *M-B* distribution function no assumption regarding the nature of energy was made so it is valid for translational, rotational, vibrational and electronic.

Later we shall prove that the quantity β appearing in the distribution function is related to the temperature T of the system through the relation

$$\beta = \frac{1}{kT} \tag{5.1.12}$$

where k is Boltzmann constant. So the M-B distribution can be written as

$$f_{MB}(\varepsilon) = e^{-\alpha} e^{-\varepsilon/kT} \tag{5.13}$$

The constant α may be expressed in terms of total number of particles N. Since

$$N = \sum n_i = e^{-\alpha} \sum_i g_i e^{-\varepsilon_i/kT}$$

Therefore

$$e^{-\alpha} = \frac{N}{\sum_i g_i e^{-\varepsilon_i/kT}} \tag{5.1.14}$$

The quantity $\sum_i g_i e^{-\varepsilon_i/kT}$ plays a fundamental role in statistical mechanics. It was first introduced by Boltzmann who called it "Zustandsumme" or sum over states. It is called *partition function Z.*

$$Z = \sum_i g_i e^{-\varepsilon_i/kT} \tag{5.1.15}$$

In terms of partition function Z the *M-B* distribution function is given by

$$n_i = \frac{N}{Z} g_i e^{-\varepsilon_i/kT}, \qquad e^{-\alpha} = \frac{N}{Z} \tag{15.1.16}$$

If the energy levels of the system are very close together, such as those of molecules in a gas, the g (ε) is replaced by g (ε) $d\varepsilon$, which represents the number of states with energies between ε and ε + $d\varepsilon$. The number of particles occupying the states in the energy range $d\varepsilon$ about ε is

$$n(\varepsilon)d\varepsilon = f_{MB}(\varepsilon)g(\varepsilon)d\varepsilon \tag{5.1.17}$$

For a system of free particles each of mass m, enclosed in a vessel of volume V, the number of states in the energy range $d\varepsilon$ about ε is

$$g(\varepsilon)d\varepsilon = g_s \cdot \frac{2\pi V}{h^3}(2m)^{3/2}\varepsilon^{1/2}\,d\varepsilon \tag{5.1.18}$$

where g_S = (2s + 1) is spin degeneracy, s = spin of the particle. For electron s = ½ , g_S = 2. For spin 0 particle g_s = 1. The function g (ε) is called *density of states.*

The expression for partition for a gas, with continuous energy levels, is obtained by replacing the summation sign by integral sign.

$$Z = \int g(\varepsilon)e^{-\beta\varepsilon}d\varepsilon \tag{5.1.19}$$

The value of parameter $e^{-\alpha}$ can be determined as follows:

$$e^{-\alpha} = \frac{N}{Z} = \frac{N}{\int g(\varepsilon)e^{-\beta\varepsilon}d\varepsilon} = \frac{N}{g_s\left(\frac{2\pi V}{h^3}(2m)^{3/2}\right)\int_0^\infty \varepsilon^{1/2}e^{-\beta\varepsilon}d\varepsilon}$$

$$= \frac{N(\beta)^{3/2}}{g_s\left(\frac{2\pi V}{h^3}(2m)^{2/3}\int_0^\infty x^{1/2}e^{-x}dx\right)} \qquad x = \beta\varepsilon, dx = \beta\, d\varepsilon$$

$$= \frac{N}{g_s\left(\frac{V}{h^3}\right)}\left(\frac{\beta}{2\pi m}\right)^{3/2} \qquad \text{where } \int_0^\infty x^{1/2}e^{-x}dx = \Gamma 3/2 = \frac{1}{2}\sqrt{\pi}$$

$$= \frac{N}{g_s V}\left(\frac{\beta h^2}{2\pi m}\right)^{3/2} \tag{5.1.20}$$

So $e^{-\alpha}$ is temperature dependent ($\beta = 1/kT$).

Substituting the value of $e^{-\alpha}$ in the expression for *M-B* distribution we have

$$f_{MB}(\varepsilon) = \frac{N}{g_s V}\left(\frac{\beta h^2}{2\pi m}\right)^{3/2} e^{-\beta\varepsilon}, \qquad \beta = \frac{1}{kT} \tag{5.1.21}$$

The number of particles occupying the energy states in the interval $d\varepsilon$ at ε in a N-particle system in equilibrium at temperature T is

$$n(\varepsilon)d\varepsilon = f_{MB}(\varepsilon)\, g(\varepsilon)d\varepsilon = \frac{2\pi N}{(\pi/\beta)^{3/2}}\varepsilon^{1/2}e^{-\beta\varepsilon}d\varepsilon, \qquad \beta = \frac{1}{kT} \tag{5.1.22}$$

Evaluation of β

The total energy of the N-particle system is

$$E = \int_0^\infty \varepsilon n(\varepsilon)d\varepsilon = \frac{2\pi N}{(\pi/\beta)^{3/2}}\int_0^\infty \varepsilon^{3/2}e^{\beta\varepsilon}d\varepsilon$$

$$= \frac{2\pi N}{(\pi/\beta)^{3/2}}\cdot\frac{1}{\beta^{5/2}}\int_0^\infty x^{3/2}e^{-x}dx, \qquad x = \beta\varepsilon, \quad dx = \beta d\varepsilon$$

$$= \frac{2N}{\sqrt{\pi}}\left(\frac{1}{\beta}\right)\Gamma(5/2)$$

$$= \frac{2N}{\sqrt{\pi}}\left(\frac{1}{\beta}\right)\left(\frac{3}{4}\sqrt{\pi}\right) = \frac{3N}{2\beta} \tag{5.1.23}$$

From kinetic theory we know that the total energy of an ideal gas containing N molecules is

$$E = \frac{3}{2}NkT \tag{5.1.24}$$

Comparison of these two expressions gives

$$\beta = \frac{1}{kT} \tag{5.1.25}$$

5.2 HEAT CAPACITY OF AN IDEAL GAS

The total energy of one mole of an ideal gas is

$$E = \frac{3}{2} N_A kT = \frac{3}{2} RT, \qquad R = N_A k = universal\ \ gas\ \ constant \qquad (5.2.1)$$

The molar heat capacity at constant volume is

$$C_v = \left(\frac{\partial E}{\partial T}\right)_v = \frac{3}{2} R \qquad (5.2.2)$$

5.3 MAXWELL'S VELOCITY DISTRIBUTION

Now we shall derive Maxwell velocity distribution for a perfect classical gas. To obtain the required distribution function we must convert the energy distribution into appropriate velocity distribution. In a perfect gas of free molecules possessing no internal degree of freedom (such as monatomic molecule) all the energy resides in the form of kinetic energy of molecules. So

$$\varepsilon = \frac{1}{2} mv^2 = \frac{1}{2} m(v_x^2 + v_y^2 + v_z^2), \qquad d\varepsilon = mvdv$$

In terms of velocity, the Maxwell distribution function becomes

$$n(v)dv = \frac{2\pi N}{(\pi kT)^{3/2}} \left(\tfrac{1}{2} mv^2\right)^{1/2} e^{-mv^2/2kT} mvdv$$

$$= 4\pi N \left(\frac{m}{2\pi kt}\right)^{3/2} v^2 e^{-mv^2/2kT} dv \qquad (5.3.1)$$

or

$$f(v)dv = \frac{n(v)}{N} = 4\pi \left(\frac{m}{2\pi kT}\right)^{3/2} v^2 e^{-mv^2/2kT} dv \qquad (5.3.2)$$

The function $f(v)$ gives the fraction of all molecules having speeds in the interval dv about v. In other words $f(v)$ represents the probability that a molecule selected at random from the gas will have its speed in the interval dv about v.

Now suppose that we wish to know how many molecules of the gas have velocities such that the x-component of velocity is in a range dv_x about v_x, the y-component is in a range dv_y about v_y, and z-component is in a range dv_z about v_z. This number of molecules is in a rectangular volume element $dv_x dv_y dv_z$ in velocity space centered on the value (v_x, v_y, v_z). We shall call this number $n(v_x, v_y, v_z)\ dv_x dv_y dv_z$.

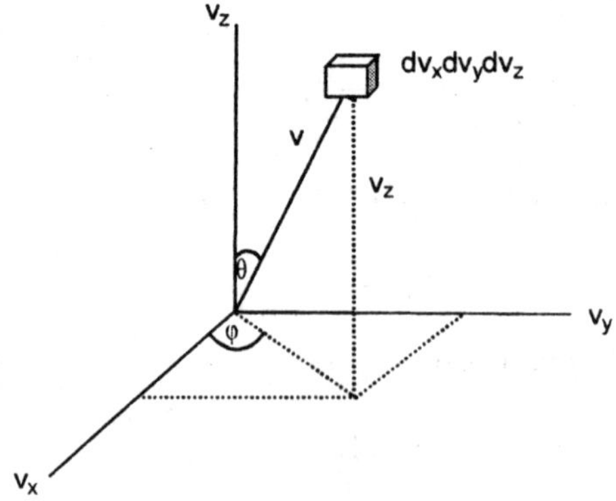

Fig. 5.3.1 Velocity space for molecules

The Maxwell-Boltzmann energy distribution function is

$$f(\varepsilon) = \frac{N}{g_s V}\left(\frac{h^2}{2\pi\, mkT}\right)^{3/2} e^{-\varepsilon/kT} \qquad (5.3.3)$$

For monatomic molecule $\quad g_s = 1.$

Therefore

$$f(v_x, v_y, v_z) = \frac{N}{g_s V}\left(\frac{h^2}{2\pi\, mkT}\right)^{3/2} e^{-m(v_x^2 + {}_y^2 + v_z^2)/2kT} \qquad (5.3.4),$$

Let us find the expression for the density of states g (v_x, v_y, v_z). We know that

$$g(p_x, p_y, p_z) dp_x dp_y dp_z = \frac{g_s V}{h^3} dp_x dp_y dp_z$$

$$g(v_x, v_y, v_z) dv_x dv_y dv_z = \frac{g_s V m^3}{h^3} dv_x dv_y dv_z \qquad (5.3.5)$$

Therefore $\quad n(v_x, v_y, v_z) dv_x dv_y dv_z = f(v_x, v_y, v_z) g(v_x, v_y, v_z) dv_x dv_y dv_z$

$$= N\left(\frac{m}{2\pi\, kT}\right)^{3/2} e^{-m(v_x^2 + v_y^2 + v_z^2)/2kT} dv_x dv_y dv_z \qquad (5.3.6)$$

Expression for *n* (v_x)

The number of molecules whose x-component of velocity lies in a range dv_x about v_x regardless of what values the y-component and z-component of velocity may have, is given by

$$n(v_x) dv_x = \int_{v_y=-\infty}^{\infty} \int_{v_z=-\infty}^{\infty} n(v_x, v_y, v_z) dv_x dv_y dv_z$$

$$= N\left(\frac{m}{2\pi\, kT}\right)^{3/2} \int_{-\infty}^{\infty} \int_{-\infty}^{\infty} e^{-m(v_x^2 + v_y^2 + v_z^2)/2kT} dv_x dv_y dv_z$$

$$= N\left(\frac{m}{2\pi\, kT}\right)^{3/2} \left(\int_{-\infty}^{\infty} e^{\frac{1}{2} m v_y^2} dv_y \int_{-\infty}^{\infty} e^{-\frac{1}{2} m v_z^2} dv_z .\right)\left(e^{-\frac{1}{2} m v_x^2} dv_x\right)$$

$$= N\left(\frac{m}{2\pi\, kT}\right)^{1/2} \left(\frac{2\pi\, kT}{m}\right)^{1/2} \left(\frac{2\pi\, kT}{m}\right)^{1/2} e^{-\frac{1}{2} m v_x^2/kT} dv_x$$

$$= N\left(\frac{m}{2\pi\, kT}\right)^{1/2} e^{-m v_x^2/2kT} dv_x \qquad (5.3.7)$$

Table of some useful definite integrals.

$$f(n) = \int_0^\infty x^n \exp(-\alpha x^2)\, dx$$

odd n	*f(n)*	*even n*	*f(n)*
1	$\frac{1}{2\alpha}$	0	$\frac{1}{2}\sqrt{\frac{\pi}{\alpha}}$
3	$\frac{1}{2\alpha^2}$	2	$\frac{1}{4}\sqrt{\frac{\pi}{\alpha^3}}$
5	$\frac{1}{\alpha^3}$	4	$\frac{3}{8}\sqrt{\frac{\pi}{\alpha^5}}$
7	$\frac{3}{\alpha^4}$	6	$\frac{15}{16}\sqrt{\frac{\pi}{\alpha^7}}$

A plot of the distribution functions $n(v)$ and $n(v_x)$ as functions of velocity coordinates is shown in the Fig. (5.3.1).

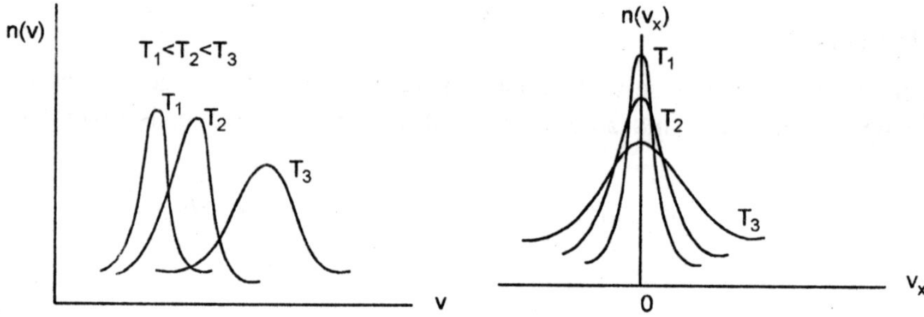

Fig. 5.3.1 Velocity distribution functions at different temperatures

Average velocity <v>

The average velocity $\langle v \rangle$ of a particle in *M-B* distribution is given by

$$\langle v \rangle = \frac{\int_0^\infty v n(v) dv}{\int_0^\infty n(v) dv} = \frac{4\pi N \left(\frac{m}{2\pi kT}\right)^{3/2} \int_0^\infty v^3 e^{-mv^2/2kT} dv}{N}$$

$$= \sqrt{\frac{8kT}{m\pi}} \tag{5.3.8}$$

Where we have used the standard result: $\int_0^\infty x^3 e^{-\alpha x^2} dx = \frac{1}{2\alpha^2}$, *where* $\alpha = \frac{m}{2kT}$.

Root mean square speed v_{rms}

The root mean square speed is defined by

$$v_{rms}^2 = \frac{1}{N}\int_0^\infty v^2\, n(v)dv = \frac{1}{N}\left[4\pi N\left(\frac{m}{2\pi kT}\right)^{3/2}\int_0^\infty v^4\, e^{-mv^2/2kT}dv\right]$$

$$= \sqrt{\frac{3kT}{m}} \tag{5.3.9}$$

Where we have used the standard result: $\int_0^\infty x^4\, e^{-\alpha x^2}dx = \frac{3}{8\alpha^2}\sqrt{\frac{\pi}{\alpha}}, \quad \alpha = \frac{m}{kT}$.

Most probable speed v_{mp}

The most probable speed corresponds to the maximum value of $n(v)$. Now

$$n(v) = 4\pi N\left(\frac{m}{2\pi kT}\right)^{3/2} v^2 e^{-mv^2/2kT}$$

The maximum value of n(v) corresponds to the speed that satisfies the equation

$$\frac{dn(v)}{dv} = 0$$

$$\frac{d}{dv}\left(v^2 e^{-mv^2/2kT}\right) = 0$$

$$v = v_{mp} = \sqrt{\frac{2kT}{m}} \tag{5.3.10}$$

Therefore

$$v_{mp} : \langle v \rangle : v_{rms} = \sqrt{\frac{2kT}{m}} : \sqrt{\frac{8kT}{m\pi}} : \sqrt{\frac{3kT}{m}}$$

$$= \sqrt{2} : \sqrt{\frac{8}{\pi}} : \sqrt{3}$$

$$= 1 : 1.13 : 1.22$$

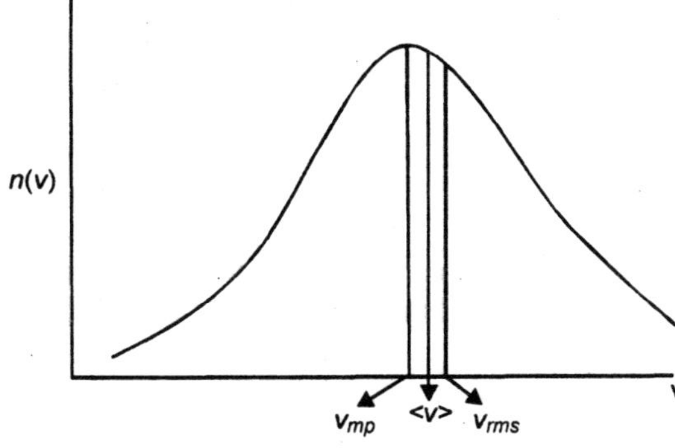

Fig. 5.3.2 Three kinds of velocities associated with Maxwell distribution

5.4 FERMI-DIRAC STATISTICS

Many properties of solid materials such as thermal and electrical conductivity, magnetic properties, specific heats etc. are related to the electron energy states. The understanding of these properties requires the distribution of electrons, which are fermions, in various states. The distribution function for electrons must satisfy two requirements: (1) electrons are indistinguishable particles *i.e.*, it is not possible to label them as electron number 1, number 2 and so on (2) electrons obey Pauli's exclusion principle, which states that no quantum state may be occupied by more than one electron. The distribution function that results from these two considerations was first developed by Fermi and Dirac in 1927. In fact all the particles, which have half odd integral spin (1/2, 3/2, 5/2 ...), obey exclusion principle, are described by *F-D* statistics. These particles are called fermions.

Consider a system composed of N fermions with total energy E. An assembly of non-interacting fermions is called fermi gas or electron gas if the fermions are electrons. The assembly of free electrons in a metal is a well-known example of fermi gas.

Let the allowed energy levels and the associated degeneracies of the system be ε_1, ε_2, ..., ε_i and g_1, g_2, ...,g_i respectively. The occupation numbers of these energy levels are n_1, n_2......n_i respectively. The most detailed description that is possible in principle is to specify the number of electrons (always 0 or 1) in each of the g_i states of energy ε_i. (Because of indistinguishability of electrons, it is meaningless to specify which electrons are in these states.) A macro state is specified by set of numbers $\{n_1, n_2, ...,n_i\}$. A microstate is specified by specifying which of the g_i states corresponding to each ε_i are occupied by electrons.

First suppose that the electrons are distinguishable. Consider a macro state $\{n_1, n_2, ...n_i\}$. Here n_i represents the number of electrons in the energy level ε_i. We want to calculate the number of ways in which n_i electrons can be placed in g_i states associated with the energy level ε_i. The first electron can be placed in any one of the g_i states. This can be done in g_i ways. For each of these choices, the second electron can be placed in $(g_i - 1)$ ways, for each of these, there are $(g_i - 2)$ ways for the third electron and so on. For the last electron there would be $g_i - (n_i-1)$ ways. The total number of possible ways of placing n_i electrons in the g_i states would be the product of all these factors. This is equal to

$$g_i(g_i-1)(g_i-2)..............(g_i-n_i+1)$$

or

$$=\frac{g_i!}{(g_i-n_i)!}$$

Since the electrons are indistinguishable, permuting the n_i electrons among the various states does not produce a physically different state of the system. There are $n_i!$ ways of permuting the particles among themselves in any given arrangement of electrons. These $n_i!$ ways don't count as separate arrangements. So we over counted the number of possible arrangement by a factor $n_i!$. The number of physically different ways of putting n_i electrons into g_i states of energy ε_i is

$$\Omega_{i,}=\frac{g_i!}{n_i!(g_i-n_i)!} \tag{5.4.1}$$

The number of microstates corresponding to a given macro state is obtained by multiplying the factors given in above expression. Thus the number Ω of micro states corresponding to the macro state specified by the set of occupation numbers $\{n_1, n_2, ...,n_i\}$ is

$$\Omega = \frac{g_1!}{n_1!(g_1-n_1)!} \cdot \frac{g_2!}{n_2!(g_2-n_2)!} \ldots\ldots = \Omega_1\Omega_2\ldots = \prod_i \Omega_i \tag{5.4.2}$$

$$\Omega = \prod_i \frac{g_i!}{n_i!(g_i-n_i)!} \tag{5.4.3}$$

g_i

| • | • | • | • | • | • | • | | |

1 2 ....n_i

| • | • | • | | | | $n_2 = 3$, $g_2 = 6$

| • | • | | | | $n_1 = 2$, $g_1 = 5$

Fig. 5.4.1

According to the fundamental principle of statistical mechanics, for a system of given total energy all the micro states are equally probable. The most probable state of the system is the most probable macro state, which corresponds to the maximum number of micro states Ω. Maximizing Ω is equivalent to maximizing $\ln \Omega$. In calculation it is more convenient to use $\ln \Omega$ than Ω. So we first calculate $\ln \Omega$.

$$\ln \Omega = \sum_i [\ln g_i! - \ln(g_i - n_i)! - \ln n_i!]$$

Using Stirling's approximation $\ln n! = n \ln n - n$ we have

$$\ln \Omega = \sum_i [g_i \ln g_i - g_i - (g_i - n_i)\ln(g_i - n_i) + (g_i - n_i) - n_i \ln n_i + n_i]$$

$$\ln \Omega = \sum_i [g_i \ln g_i - (g_i - n_i)\ln(g_i - n_i) - n_i \ln n_i] \tag{5.4.4}$$

The most probable distribution is one for which a small change δn_i in any of n_i has no effect on the value of Ω. We assume that n_i are continuous, so the condition of maximum $\ln \Omega$ becomes

$$\frac{\partial \ln \Omega}{\partial n_i} = 0 \text{ for each } n_i.$$

If the change in $\ln \Omega$ corresponding to change δn_i in n_i is $\delta \ln \Omega$ then

$$\delta \ln \Omega = \sum_i [\ln(g_i - n_i) - \ln n_i]\delta n_i = 0 \tag{5.4.5}$$

Note that n_i are not independent, but they are related through the conditions that the total number of particle is constant.

$$\Sigma\, n_i = N = \text{constant.} \tag{5.4.6}$$

And the total energy of the system is constant.

$$\Sigma\, n_i\, \varepsilon_i = E = \text{constant.} \tag{5.4.7}$$

From these two equations we have

$$\Sigma\, \delta n_i = 0 \tag{5.4.8}$$

and $$\Sigma\, \varepsilon_i\, \delta\, n_i = 0 \tag{5.4.9}$$

In order to incorporate the conditions of conservation of number of particles and of energy in Eq. (5.4.5) we use Lagrange method of undetermined multipliers. To do so, we multiply Eq. (5.4.8) by $-\alpha$ and Eq. (5.4.9) by $-\beta$ and add them to Eq. (5.4.5). Here α and β are independent of n_i.

$$\sum_i [\ln(g_i - n_i) - \ln n_i - \alpha - \beta\,\varepsilon_i]\delta\, n_i = 0$$

$$\ln(g_i - n_i) - \ln n_i - \alpha - \beta\varepsilon_i = 0$$

$$n_i = \frac{g_i}{e^{\alpha + \beta\varepsilon_i} + 1} \tag{5.4.10}$$

The ration n_i/g_i is called *F–D* distribution function and is denoted by f_{FD}.

$$f_{FD}(\varepsilon) = \frac{n_i}{g_i} = \frac{1}{e^{\alpha}\, e^{\beta\,\varepsilon} + 1} \tag{5.4.11}$$

The function f_{FD} (ε) is the average number of electrons per quantum state of energy ε. It also represents the probability that a state of energy ε is occupied. The quantity β is $1/kT$, where T is temperature of the system. The value of α is determined by the normalization condition

$$\sum_i n_i = N = \sum_i \frac{g_i}{e^{\alpha}\ \ e^{\beta\,\varepsilon_i} + 1}$$

It is customary to express α in terms of another constant μ, called chemical potential or E_F, called Fermi energy as follows.

$$\alpha = -\beta\mu = -\beta\,\varepsilon_F \quad or \quad \alpha = -\frac{\mu}{kT} = -\frac{\varepsilon_F}{kT} \tag{5.4.12}$$

In terms of chemical potential μ and Fermi energy ε_F the *FD* distribution becomes

$$f_{FD}(\varepsilon) = \frac{1}{e^{\beta(\varepsilon-\mu)} + 1} = \frac{1}{e^{(\varepsilon-\varepsilon_F)/kT} + 1} \tag{5.4.13}$$

For systems with continuous energy levels, the number of fermions occupying the states in the energy range $d\varepsilon$ at ε is given by

$$n(\varepsilon)d\varepsilon = g(\varepsilon) f_{FD}(\varepsilon) d\varepsilon \tag{5.4.14}$$

The total number of fermion in the system is given by

$$N = \int n(\varepsilon)d\varepsilon = \int g(\varepsilon) f_{FD}(\varepsilon)d\varepsilon \tag{5.4.15}$$

For electron gas, the density of states is given by

$$g(\varepsilon)d\varepsilon = (2s+1)\frac{2\pi\, V}{h^3}(2m)^{3/2}\,\varepsilon^{1/2}\, d\varepsilon, \qquad s = 1/2 \quad for\ electron. \tag{5.4.16}$$

5.5 BOSE–EINSTEIN STATISTICS

There are many systems, which are composed of weakly interacting identical and indistinguishable particles with integral spin. These particles don't obey Pauli's exclusion principle. So any number of particles can occupy the same quantum state. The statistical behavior of such particles is governed by a different kind of statistics, called Bose-Einstein statistics, named in honor of S.N. Bose and A. Einstein who independently derived it. Particles obeying Bose-Einstein statistics are called *bosons*. Examples of *bosons* are photons (spin 1), H_2 molecule, helium 4_{He}, meson etc.

Consider a system composed of N bosons. The macrostate state of the system is specified by specifying the occupation numbers n_1, n_2, ...in energy levels ε_1, ε_2, ... having degeneracies g_1, g_2 ....The *i-th* level with energy ε_i and degeneracy g_i contains n_i particles. There is no restriction on the number of bosons that a quantum state can accommodate. This level can be pictured as n_i particles in a row divided arbitrarily into g_i states by $g_i - 1$ partitions.

```
           O O | O |    | O O O O | O O |      | O | O O O | O
g =          1   2    3      4        5     6   7     8     9
n_i = 14
```

Fig. 5.5.1 Distribution of 14 particles in 9 states separated by 8 partitions

The number of different ways Ω_i, the n_i bosons can be placed in the g_i quantum states without any limit to the number of particles in a state is equal to the number of independent permutations of particles and partitions. There are a total of $(n_i + g_i - 1)$ particles plus partitions, which can be arranged in $(n_i + g_i - 1)!$ ways. Since the permutation of particles among themselves and permutations of partitions among themselves don't produce a different arrangement, we must, therefore, divide $(n_i + g_i - 1)!$ by $n_i! \cdot (g_i - 1)!$. Thus

$$\Omega_i = \frac{(n_i + g_i - 1)!}{n_i!\,(g_i - 1)!} \tag{5.5.1}$$

The total number Ω of different ways to arrange n_1, n_2, ... bosons in the energy levels ε_1, ε_2, ..., if there are g_1, g_2 ..... states in each level, is

$$\Omega = \frac{(n_1 + g_1 - 1)!}{n_1!(g_1 - 1)!} \cdot \frac{(n_2 + g_2 - 1)!}{n_2!(g_2 - 1)!} \cdots\cdots\cdots$$

$$= \prod_i \frac{(n_i + g_i - 1)!}{n_i!(g_i - 1)!} \tag{5.5.2}$$

Since $n_i >> 1$, $g_i >> 1$, 1 may be omitted in the above expression. Thus

$$\Omega = \prod_i \frac{(n_i + g_i)!}{n_i!\, g_i!} \tag{5.5.3}$$

If the number N of the particles and the total energy E of the system is constant then we have

$$n_1 + n_2 + \ldots\ldots = \sum_i n_i = N \tag{5.5.4}$$

$$n_1\varepsilon_1 + n_2\varepsilon_2 \ldots\ldots\ldots = \sum_i n_i\varepsilon_i = E \tag{5.5.5}$$

Since N and E are constants, the sum of changes in occupation numbers and the sum of changes in energies of the energy levels must be zero.

$$\sum_i \delta n_1 = 0 \tag{5.5.6}$$

$$\sum_i \varepsilon_i \delta n_i = 0 \tag{5.5.7}$$

The most probable distribution corresponds to the maximum value of Ω subject to the restrictions expressed by Eq. (5.5.6) and (5.5.7). It is more convenient to maximize ln Ω than Ω. So we first simplify ln Ω.

$$\ln \Omega = \sum \ln (n_i + g_i)! - \ln n_i! - \ln g_i!$$

Using Stirling approximation we have

$$\ln \Omega = \sum \left[(n_i + g_i)\ln(n_i + g_i) - (n_i + g_i) - n_i \ln n_i + n_i - g_i \ln g_i + g_i\right]$$

$$= \sum \left[(n_i + g_i)\ln(n_i + g_i) - n_i \ln n_i - g_i \ln g_i\right] \tag{5.5.8}$$

The condition of maximum ln Ω is

$$\delta \ln \Omega = 0$$

$$\sum \left[\ln(n_i + g_i) + (n_i + g_i)\cdot\frac{1}{(n_i + g_i)} - \ln n_i - n_i \cdot \frac{1}{n_i}\right] \delta\, n_i = 0$$

$$\sum \left[\ln(n_i + g_i) - \ln n_i\right] \delta\, n_i = 0 \tag{5.5.9}$$

To incorporate the conditions (5.5.6) and (5.5.7) we use Lagrange method of undetermined multipliers. Multiplying (5.5.6) by – α and (5.5.7) by – β and adding them to Eq.(5.5.9) we have

$$\sum \left[\ln(n_i + g_i) - \ln n_i - \alpha - \beta \varepsilon_i\right] \delta n_i = 0$$

Since the change δn_i's are arbitrary, we must have

$$\ln(n_i + g_i) - \ln n_i - \alpha - \beta \varepsilon_i = 0$$

This simplifies to

$$n_i = \frac{g_i}{e^{\alpha + \beta \varepsilon_i} - 1} \tag{5.5.10}$$

$$f_{BE}(\varepsilon_i) = \frac{n_i}{g_i} = \frac{1}{e^{\alpha + \beta \varepsilon_i} - 1}, \quad \beta = 1/kT \tag{5.5.11}$$

Eq. (5.5.10) represents the *BE* distribution function. Eq.(5.5.11) represents number of bosons per quantum state at energy ε_i or the occupation probability of state with energy ε_i of a system in thermal equilibrium at temperature *T*.

In a system in which the number of bosons is not conserved, the condition $\Sigma\, n_i = N =$ constant does not apply. For example the number photons in a cavity increases with increasing temperature. This in contrast to an ideal gas contained in a vessel. Removal of this condition is equivalent to setting α = 0. For such system the *B-E* distribution becomes

$$n_i = \frac{g_i}{e^{\beta \varepsilon_i} - 1}$$

$$f_{BE}(\varepsilon) = \frac{1}{e^{\beta\varepsilon} - 1} \qquad \text{(for photons } \varepsilon = \hbar\omega) \tag{5.5.12}$$

The parameter α may be determined from the condition $\Sigma\, n_i = N$. It increases monotonically with temperature for *FD* and *BE* statistics both. This can be seen as follows. The number of bosons occupying the states with energy between ε and ε + dε is given by

$$n(\varepsilon)d\varepsilon = f_{BE}(\varepsilon)g(\varepsilon)d\varepsilon \tag{5.5.13}$$

The total number of bosons in the system is

$$N = \int_0^\infty f_{BE}(\varepsilon)g(\varepsilon)d\varepsilon$$

$$= \frac{2\pi V(2m)^{3/2}}{h^3}\int_0^\infty \frac{\varepsilon^{1/2}\, d\varepsilon}{e^\alpha . e^{\varepsilon/kT} - 1} \tag{5.5.14}$$

Putting ε/kT = q, in the above integral we have

$$N = \frac{2\pi V(2mkT)^{3/2}}{h^3}\int_0^\infty \frac{q^{1/2}\, dq}{e^\alpha e^q - 1} \tag{5.5.15}$$

With increasing *T*, the factor multiplying the integral in the above expression (5.5.15) increases. Since *N* is constant, the integral must decease. This implies that α increases with rise in temperature. Thus α is an increasing function of *T*. Since *N* is finite, the integral must always converge and so α must always be non-negative.

6

Applications of Quantum Statistics

6.1 APPLICATIONS OF FERMI–DIRAC STATISTICS

6.1.1 Sommerfeld's Free Electron Theory of Metals

According to free electron model, the valence electrons of atoms constituting the metal are free to move within the limits of the metal. The potential energy of interaction of electrons with the ion cores is constant throughout the solid and may be assumed to be zero for convenience. The free electrons don't leave the boundaries of the metal because of the electrostatic force. The potential energy of electrons may be assumed to be infinitely great at the boundaries. Thus the electrons in a metal may be treated as a gas which is composed of non-interacting spin ½ fermions confined in box. Because of this analogy the assembly of free electrons in a metal is called Fermi gas. Quantum mechanical treatment of motion of electron in a box shows that energy of electron is quantized. The energy of electron, which is free to move in a cubical box of side L is given by

$$\varepsilon = \frac{\pi^2 \hbar^2}{2mL^2}\left(n_x^2 + n_y^2 + n_z^2\right) \tag{6.1.1}$$

where n_x, n_y, n_z are integers, each can take on values 1,2, ...The set of integers n_x, n_y, n_z and spin quantum number ½ define a state of electron. It is found that more than one quantum state correspond to a single energy level. The different quantum states belonging to an energy level are called *degenerate states* and the number of such states is called the *degeneracy* of that energy level.

The density of states (the number of quantum states per unit energy interval) at energy ε is given by

$$g(\varepsilon) = (2s + 1)\frac{2\pi V}{h^3}(2m)^{3/2}\ \varepsilon^{1/2}. \tag{6.1.2}$$

For electron spin s = ½ , and $(2s + 1) = 2$. The free electrons in a metal are distributed among various available quantum states according to Pauli's exclusion principle. The Fermi-Dirac distribution function gives the probability that a state with energy ε is occupied at temperature T.

Temperature dependence of *F-D* Distribution and Fermi Energy $\varepsilon_{F(0)}$

F-D distribution is

$$f_{FD}(\varepsilon) = \frac{1}{e^{(\varepsilon - \varepsilon_F)/kT} + 1} \tag{6.1.3}$$

At $T = 0$ K, for the energy states with $\varepsilon < \varepsilon_{F0}$ where ε_{F0} is the Fermi energy at 0 K, the exponential term

$$\exp(\varepsilon - \varepsilon_F)/kT \rightarrow \exp(-\infty) \rightarrow 0 \text{ and therefore}$$

$$f_{FD} = 1$$

At $T = 0$ K, for energy states with $\varepsilon > \varepsilon_F$, the exponential term $\exp(\varepsilon - \varepsilon_F)/kT \rightarrow \exp(\infty) \rightarrow \infty$. So

$$f_{FD} = 0$$

Thus at $T = 0$ K, all the energy states below the Fermi level are occupied and those above it are empty. Fermi energy is the maximum value of energy that a fermion can acquire at 0 K.

At $T \neq 0$ K the probability the energy level $\varepsilon = \varepsilon_F$ is occupied is

$$f_{FD}(\varepsilon = \varepsilon_F) = \frac{1}{e^0 + 1} = \frac{1}{2}$$

That is, at Fermi energy one half the energy states will be occupied. The variation of FD distribution with energy at different temperatures is shown in the figure (6.1.1).

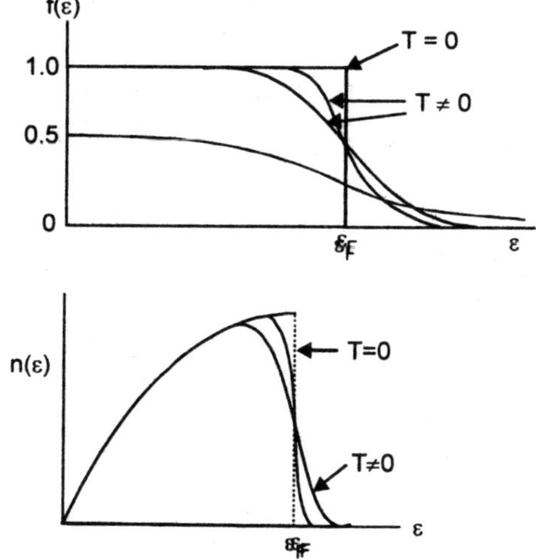

Fig. 6.1.1 Dependence of Fermi distribution function f_{FD} and number of electrons with temperature

From the figure it is evident that at temperature T above 0 K, some of the states slightly below ε_F are empty and some of the states slightly above the ε_F are full. The explanation for this is that as the temperature of the metal is raised, a small fraction of the electrons occupying the states just below the fermi level gain thermal energy and get excited to the states just above the fermi level.

For a free electron gas the number of states in the energy interval $d\varepsilon$ at ε is

$$g(\varepsilon)d\varepsilon = \frac{4\pi V}{h^3}(2m)^{3/2}\, \varepsilon^{1/2}\, d\varepsilon \tag{6.1.4}$$

The number of electrons occupying the energy states between the energy interval $d\varepsilon$ at ε is

$$n(\varepsilon)d\varepsilon = f_{FD}(\varepsilon)\, g(\varepsilon)d\varepsilon.$$

The Fermi energy is in general a function of temperature. Its value is determined by the condition

$$N = \int n(\varepsilon)d\varepsilon = \int f_{FD}(\varepsilon)g(\varepsilon)d\varepsilon = \int \frac{g(\varepsilon)}{1+\exp(\varepsilon - \varepsilon_F)/kT} d\varepsilon$$

where the integral is taken over all the energy states available to the electrons of the system. Substituting the value of g (ε) in above expression we have

$$N = \frac{4\pi V}{h^3}(2m)^{3/2} \int_0^\infty \frac{\varepsilon^{1/2}}{1+\exp(\varepsilon - \varepsilon_F)/kT} d\varepsilon \tag{6.1.6}$$

At T = 0K,

$$f_{FD} = 1 \quad \text{for} \quad \varepsilon < \varepsilon_{F0}$$
$$= 0 \quad \text{for} \quad \varepsilon > \varepsilon_{F0}$$

So the limits of integration can be taken from 0 to ε_{F0}. Then

$$N = \frac{4\pi V(2m)^{3/2}}{h^3} \int_0^{\varepsilon_{F0}} \varepsilon^{1/2} d\varepsilon \qquad (f_{FD} = 1)$$

$$= \frac{4\pi V(2m)^{3/2}}{h^3}\left(\frac{2}{3}\varepsilon_{F0}^{3/2}\right)$$

$$\varepsilon_{F0} = \frac{h^2}{8m}\left(\frac{3N}{\pi V}\right)^{2/3} \tag{6.1.7}$$

$$= \left(3.646 \times 10^{-19}\, eV.m^2\right)\left(\frac{N}{V}\right)^{2/3} \tag{6.1.8}$$

For copper N/V = 8.5×10^{28} m^{-3}, the Fermi energy is

$$\varepsilon_{F0} = (3.656 \times 10^{-19} \text{eV.m}^2)(8.5 \times 10^{28} \text{m}^{-3})^{2/3} = 7.0 \text{ eV}.$$

The energy of an electron confined to move in a cubical box of side L is

$$\varepsilon = \frac{n^2 h^2}{8mL^2}$$

The quantum number of electron occupying the highest energy state $\varepsilon = \varepsilon_{F0}$ is

$$n_{\max} = \frac{\sqrt{8m\varepsilon_{F0}}}{h} L$$

For a box of size L = 1 cm, $n_{\max} = 43 \times 10^6$. Thus we see that the quantum numbers of the occupied states may from 1 to 43 million. The existence of such a huge number of states allows us to treat the energy levels as continuous.

In order to express density of states $g(\varepsilon)$ in terms of Fermi energy ε_{F0} we multiply the expressions for $g(\varepsilon)$ and ε_{F0}. Doing so we get

$$g(\varepsilon) = \frac{3N}{2\varepsilon_{F0}^{3/2}}\varepsilon^{1/2} \tag{6.1.9}$$

For 1 mol of copper, $N = N_A = 6.02 \times 10^{23}\ \text{mol}^{-1}$, $\varepsilon_{F0} = 7\text{eV}$, we have

$$g(\varepsilon_{F0}) = \frac{3\times 6.02\times 10^{23}}{2(7\text{eV})^{3/2}}(7\text{eV})^{1/2} = 1.3\times 10^{23}\ \text{states/eV}$$

With this huge number of states per unit energy range, it is clear that we may consider the energy to be virtually continuous.

Average Kinetic Energy per Electron at 0 K

The average energy of electrons at 0 K is

$$\langle\varepsilon\rangle = \frac{1}{N}\int \varepsilon\, n(\varepsilon)d\varepsilon = \frac{1}{N}\int \varepsilon\, f_{FD}(\varepsilon)\, g(\varepsilon)\, d\varepsilon$$

$$= \frac{4\pi V}{Nh^3}(2m)^{3/2}\int_0^\infty \frac{\varepsilon^{3/2}\, d\varepsilon}{1 + \exp(\varepsilon - \varepsilon_F)/kT}$$

$$= \frac{4\pi V}{Nh^3}(2m)^{3/2}\int_0^{\varepsilon_{F0}} \varepsilon^{3/2}\, d\varepsilon$$

(At $T = 0$ K, $f_{FD} = 1$ for $\varepsilon < \varepsilon_{F0}$)

$$= \frac{4\pi V}{Nh^3}(2m)^{3/2}\left(\frac{2}{5}\varepsilon_{F0}^{5/2}\right) \tag{6.1.10}$$

Making use of the result

$$N = \frac{4\pi V}{h^3}(2m)^{3/2}\left(\frac{2}{3}\varepsilon_{F0}^{3/2}\right) \tag{6.1.11}$$

we can express $\langle\varepsilon\rangle$ as

$$\langle\varepsilon_0\rangle = \frac{3}{5}\varepsilon_{F0} \tag{6.1.12}$$

The total energy of the system is

$$U(0) = \frac{3}{5}N\varepsilon_{F0} \tag{6.1.13}$$

Fermi Temperature T_F

The Fermi temperature T_F of a system is defined as

$$T_F = \frac{\varepsilon_{F0}}{k}, \text{ where } k \text{ is Boltzmann constant.} \tag{6.1.14}$$

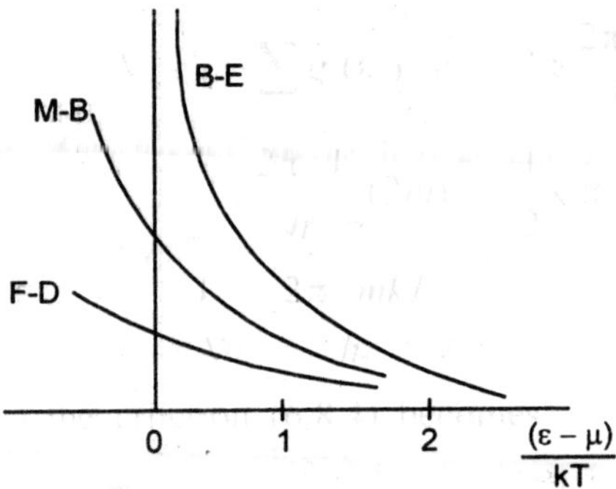

Fig. 6.1.2 Variation of statistical distribution function with (ε–M)//*kT*.

For $T << T_F$ or $kT_F << \varepsilon_{F0}$ the *F-D* distribution is called degenerate and for $T >> T_F$ the distribution is non-degenerate. The parameter $\alpha = -\mu/kT$ is negative for degenerate systems and positive for non-degenerate systems. This means that $\mu > 0$ at low temperature and $\mu < 0$ at high temperature.

At $T = 0$ *K* the system is completely degenerate, at $T << T_F$ the system is degenerate and at intermediate temperature it is slightly degenerate.

Table 6.1 Fermi energy, Fermi temperature and Fermi velocity of some metals

Element	*N/V cm*$^{-3}$ (10^{22})	ε_F *eV*	T_F *K* (10^4)	v_F *cm/s* (10^8)
Li	4.6	4.7	5.5	1.3
K	1.34	2.1	2.4	0.85
Cu	8.5	7.0	8.2	1.56
Au	5.9	5.5	6.4	1.39

Dependence of Fermi Energy (Chemical Potential) with Temperature

The total number of electrons in a Fermi gas is

$$N = \int_0^\infty f_{FD}(\varepsilon)g(\varepsilon)d\varepsilon = \frac{4\pi V}{h^3}(2m)^{3/2}\int_0^\infty \frac{\varepsilon^{1/2}}{1+\exp(\varepsilon-\varepsilon_F)/kT}d\varepsilon$$

$$= C.\frac{2}{3}\left[\varepsilon_F^{3/2} + 2(kT)^2\left(1-\frac{1}{2}\right)\zeta(2)\left(\frac{3}{4}\varepsilon_F^{-1/2}\right)\right]$$

$$= \frac{2}{3}C\varepsilon_F^{3/2}\left[1+\frac{\pi^2}{8\varepsilon_F^2}(kT)^2\right] \quad (6.1.15)$$

where $$C = \frac{4\pi V}{h^3}(2m)^{3/2} \quad (6.1.16)$$

In the limit $T \to 0$

$$N = \frac{2}{3} C \varepsilon_{F0}^{3/2} \tag{6.1.17}$$

Therefore
$$\varepsilon_{F0} = \left(\frac{3N}{2C}\right)^{2/3} = \frac{h^2}{8m}\left(\frac{3N}{\pi V}\right)^{2/3} \tag{6.1.18}$$

Since kT/ε_F is small, we see that ε_F does not change rapidly with temperature. Therefore, we can set $\varepsilon_F = \varepsilon_{F0}$ in the second term on the right hand side of Eq. (6.1.15) to obtain, after putting

$N = \frac{2}{3} C \varepsilon_{F0}^{3/2}$. Then we have

$$\frac{2}{3} C \varepsilon_{F0}^{3/2} = \frac{2}{3} C \varepsilon_F^{3/2} \left[1 + \frac{\pi^2}{8}\left(\frac{kT}{\varepsilon_{F0}}\right)^2\right]$$

From this we get

$$\varepsilon_F = \varepsilon_{F0}\left[1 + \frac{\pi^2}{8}\left(\frac{kT}{\varepsilon_{F0}}\right)^2\right]^{-2/3} \tag{6.1.19}$$

Using the result $(1 + x)^{-2/3} = 1 - 2x/3$ we can write

$$\varepsilon_F = \varepsilon_{F0}\left[1 - \frac{\pi^2}{12}\left(\frac{kT}{\varepsilon_{F0}}\right)^2\right] \tag{6.1.20}$$

Fig (6.1.3) Variation of chemical potential μ with temperature T for a gas of non-interacting particles at fixed density. T_C is Bose-Einstein condensation temperature. For fermions, at $T < T_F$, μ is positive, (α is negative) and at $T > T_F$, μ is negative (α positive).

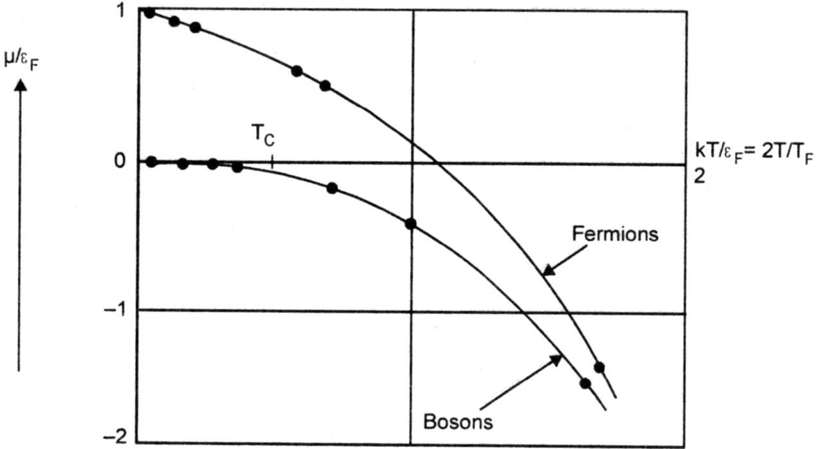

Fig. 6.1.3 Variation of chemical potential with temperature.

Total Energy of Electron Gas at Low Temperature

The total energy of electron gas is given by

$$E = \int_0^\infty \varepsilon\, f_{FD}(\varepsilon) g(\varepsilon) d\varepsilon = C \int_0^\infty \frac{\varepsilon^{3/2}}{1+\exp(\varepsilon - \varepsilon_F)/kT} d\varepsilon \quad \text{where} \quad g(\varepsilon) = C\varepsilon^{1/2}$$

$$= \frac{2C}{5}\left[\varepsilon_F^{5/2} + 2(kT)^2 (1/2)(\pi^2/6)\left(\frac{15}{4}\varepsilon_F^{1/2}\right)\right]$$

$$= \frac{2C}{5}\varepsilon_F^{5/2}\left[1 + \frac{5\pi^2}{8}\left(\frac{kT}{\varepsilon_F}\right)^2\right] \tag{6.1.21}$$

Substituting $C = \frac{3}{2} N \varepsilon_{F0}^{-3/2}$ and ε_F and recalling that ε_F does not change rapidly with temperature we can express the total energy of electron gas as

$$E = \frac{2}{5}\left[\frac{3}{2} N\varepsilon_{F0}^{-3/2}\right]\left[\varepsilon_{F0}^{5/2}\left\{1 - \frac{\pi^2}{12}\left(\frac{kT}{\varepsilon_{F0}}\right)^2\right\}\right]\left[1 + \frac{5\pi^2}{8}\left(\frac{kT}{\varepsilon_{F0}}\right)^2\right]$$

$$\approx \frac{3}{5} N\varepsilon_{F0}\left[1 + \frac{5\pi^2}{12}\left(\frac{kT}{\varepsilon_{F0}}\right)^2 - \frac{\pi^2}{16}\left(\frac{kT}{\varepsilon_F}\right)^4 + ...\right] \tag{6.1.22}$$

6.2 ELECTRONIC HEAT CAPACITY

According to classical theory, when a system of particles is heated, all the particles absorb heat and contribute to the heat capacity. Thus the classical theory applied to electron gas in a monovalent metal predicts electronic contribution to heat capacity equal to $3R/2$. But experimental results are found to be less than 1% of the classical value. This anomaly is removed if one uses Fermi-Dirac distribution function to the electron gas. When a metal is heated only a small fraction of electrons, which are within an energy kT below the Fermi level are excited thermally to vacant states above the Fermi level. Electrons, which are deeply situated below the Fermi level don't participate in thermal excitation because the energy kT is not enough to take them to the vacant levels above the Fermi level. This explains why electronic contribution to heat capacity is very small.

The total energy of electron gas is

$$E \approx \frac{3}{5} N\varepsilon_{F0}\left[1 + \frac{5\pi^2}{12}\left(\frac{kT}{\varepsilon_{F0}}\right)^2\right] \tag{6.2.1}$$

The heat capacity of electron gas is

$$C_{ve} = \frac{\partial E}{\partial T} = \left(\frac{N\pi^2 k^2}{2\varepsilon_{F0}}\right) T \tag{6.2.2}$$

Let us calculate the value of C_{ve} for copper. For copper ε_{F0} = 7 eV. Substituting $N_A = 6.02\times10^{26}$ (kmol)$^{-1}$, $k = 1.38\times10^{-23}$JK^{-1}, Room temperature T = 300 K, kT = 0.026 eV.

$$C_{ve} = \frac{\pi^2}{2} \cdot \frac{kT}{\varepsilon_{F0}} \cdot kN_A = \frac{(3.14)^2}{2} \cdot \frac{0.026\ eV}{7.0\ eV} \cdot (1.38 \times 10^{-23}\ \text{J/K})\ (6.02 \times 10^{26}\ \text{kmol}^{-1})$$

$$= 1460\ \text{J (kmol)}^{-1}\ \text{K}^{-1}.$$

We know that at very low temperature, the lattice heat capacity varies as T^3 (Debye T^3 law) while the electronic heat capacity varies linearly with temperature T. At very low temperature the lattice heat capacity decreases very rapidly and the electronic heat capacity dominates. At high temperature the lattice heat capacity dominates over the electronic contribution. The total heat capacity is given by

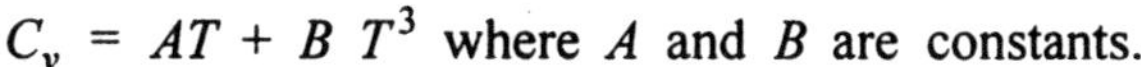

$$C_v = AT + B\,T^3 \text{ where } A \text{ and } B \text{ are constants.}$$

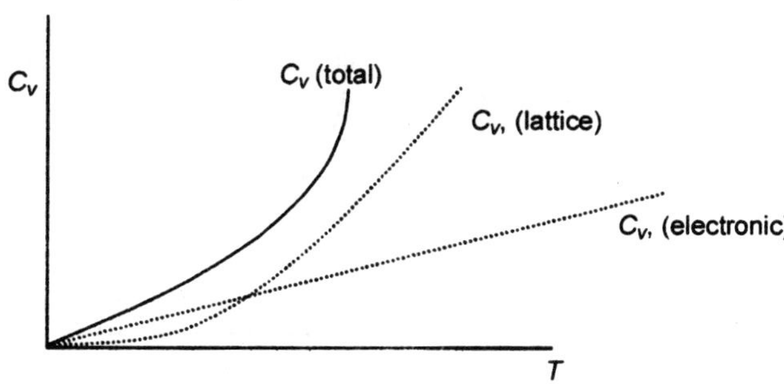

Fig. 6.2.1 Variation of heat capacity with temperature.

6.3 THERMIONIC EMISSION (RICHARDSON-DUSHMANN EQUATION)

The emission of electrons from a substance when it is heated to a high temperature is called *thermionic emission*. The thermionic current density depends on the temperature T, work function and the nature of the emitting surface. The expression, which represents the dependence of thermionic current on temperature and work function of emitting material, was first derived by Richardson making use thermodynamic principle and later by Dushmann using quantum statistics developed by Fermi and Dirac.

According free electron model, the electron are distributed among various available quantum states according to Pauli's exclusion principle. At $T = 0\ K$, all the states up to fermi level ε_F are filled and those above it are empty. The work function φ denotes the energy required to liberate the electron at Fermi level from the metal. In order to liberate electron from a metal, the energy ε imparted to it must exceed $(\varepsilon_F + \varphi)$. For an electron to escape, it must arrive at the emitting surface with momentum p suitably directed. We take the emitting material in the shape of a rectangular box with emitting surface perpendicular to x-axis. For electron emission to take place, the x-component of momentum p_x must be greater than the critical value p_{xc} given by

$$p_x \geq p_{xc} = \sqrt{2m(\varepsilon_F + \varphi)} \tag{6.3.1}$$

Let $n(p_x)dp_x$ represent the number of electrons per unit volume having x-component of momentum in the range dp_x at p_x. The thermionic current density J is given by

$$J = \int_{p_{xc}}^{\infty} n(p_x)dp_x.e.v_x = \frac{e}{m}\int_{p_{xc}}^{\infty} p_x.n(p_x)dp_x \tag{6.3.2}$$

where v_x is x-component of velocity of electron.

The number of quantum states in volume element $dxdydzdp_xdp_ydp_z$ of phase space is

$\frac{2}{h^3}dxdydzdp_xdp_ydp_z$. The presence of factor 2 accounts for the fact that for a given momentum state p_x, p_y, p_z, there can be two spin states: spin up and spin down. The number of quantum states in unit volume of coordinate space and in volume element $dp_x\,dp_y\,dp_z$ of momentum space is denoted by $g(p_x, p_y, p_z)dp_xdp_ydp_z$ and will be given by

$$g\,(p_x, p_y, p_z)dp_xdp_ydp_z = 2\frac{dp_xdp_ydp_z}{h^3}.$$

The number of electrons per unit volume with momentum between p_x and $p_x + dp_x$, p_y and $p_y + dp_y$, p_z and $p_z + dp_z$ is

$$n(p_x,p_y,p_z)dp_xdp_ydp_z = 2\frac{dp_xdp_ydp_z}{h^3}.f_{FD}(\varepsilon)$$

$$= 2\frac{1}{1+\exp[(\varepsilon - \varepsilon_F)/kT]}\cdot\frac{dp_xdp_ydp_z}{h^3} \tag{6.3.3}$$

where $\varepsilon = (p_x^2 + p_y^2 + p_z^2)/2m$ is the energy of electron.

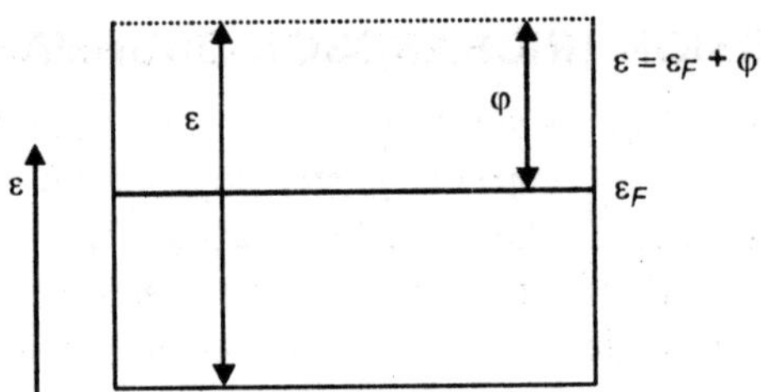

Fig. 6.3.1 Energy of electron ε in terms of fermi energy ε_F and work function ϕ.

The number of electrons with x-component of momentum between p_x and $p_x + dp_x$, irrespective of the values that p_y and p_z can assume, is given by

$$n(p_x)dp_x = \frac{2}{h^3}dp_x\int_{-\infty}^{\infty}\int_{-\infty}^{\infty}\frac{1}{1+\exp(\varepsilon - \varepsilon_F)/kT}dp_ydp_z \tag{6.3.4}$$

At any temperature $(\varepsilon - \varepsilon_F)/kT >> 1$ so 1 may be neglected in the denominator of the *F-D* distribution. Under this approximation we have

$$n(p_x)dp_x = \frac{2}{h^3}dp_x\int_{-\infty}^{\infty}\int_{-\infty}^{\infty}\left[\exp\left(-\frac{p_x^2+p_y^2+p_z^2}{2m}+\varepsilon_F\right)/kT\right]dp_ydp_z$$

$$= \frac{2}{h^3}\exp\left(\frac{\varepsilon_F}{kT}\right).\exp\left(-\frac{p_x^2}{2mkT}\right).dp_x.\int_{-\infty}^{\infty}\exp\left(-\frac{p_y^2}{2mkT}\right)dp_y.\int_{-\infty}^{\infty}\exp\left(-\frac{p_z^2}{2mkT}\right)dp_z$$

$$= \frac{2}{h^3} \exp\left(\frac{\varepsilon_F}{kT}\right) . \exp\left(-\frac{p_x^2}{2mkT}\right) . \left(\sqrt{2\pi mkT}\right)\left(\sqrt{2\pi mkT}\right) dp_x \tag{6.3.5}$$

where we have used the standard result: $\int_0^{\infty} \exp(-\alpha x^2) dx = \frac{1}{2}\sqrt{\frac{\pi}{\alpha}}$. In view of this result we have

$$n(p_x) dp_x = \frac{4\pi mkT}{h^3} . \exp\left(\frac{\varepsilon_F}{kT}\right) . \exp\left(-\frac{p_x^2}{2mkT}\right) dp_x \tag{6.3.6}$$

The thermionic current now becomes

$$J = \frac{e}{m} . \frac{4\pi mkT}{h^3} . \exp\left(\frac{\varepsilon_F}{kT}\right) . \int_{p_{xc}}^{\infty} p_x . \exp\left(-\frac{p_x^2}{2mkT}\right) . dp_x$$

$$= \left(\frac{4\pi mek^2}{h^3}\right) T^2 \exp\left(\frac{\varepsilon_F - \dfrac{p_{xc}^2}{2m}}{kT}\right)$$

$$= \left(\frac{4\pi mek^2}{h^3}\right) T^2 e^{-\varphi / kT}$$

$$= AT^2 \exp(-\varphi / kT) \tag{6.3.7}$$

where $$A = \frac{4\pi mek^2}{h^3} = 1.20 \times 10^6 \, Am^{-2} K^{-2} \tag{6.3.8}$$

From above relation

$$\ln \frac{J}{T^2} = \ln A - \frac{\varphi}{kT}$$

$$\log_{10} \frac{J}{T^2} = \log_{10} A - 0.434 \frac{\varphi}{kT}$$

A plot of $\log_{10} J/T^2$ against $1/T$ is a straight line. The intercept on y-axis gives $\log_{10} A$ and the slope of the line gives the work function of the emitting material.

The value of A determined from experiment does not agree with the theoretical value obtained by putting the values of constants in the expression for A. A correction needs in the expression for the thermionic current density. When an electron leaves the emitting surface, the latter becomes positive and pulls the outgoing electron back to the material. If r denotes the fraction of electrons reflected back to the material, the thermionic current density will be given by

$$J = (1-r)AT^2 \exp\left(-\frac{\varphi}{kT}\right) \tag{6.3.9}$$

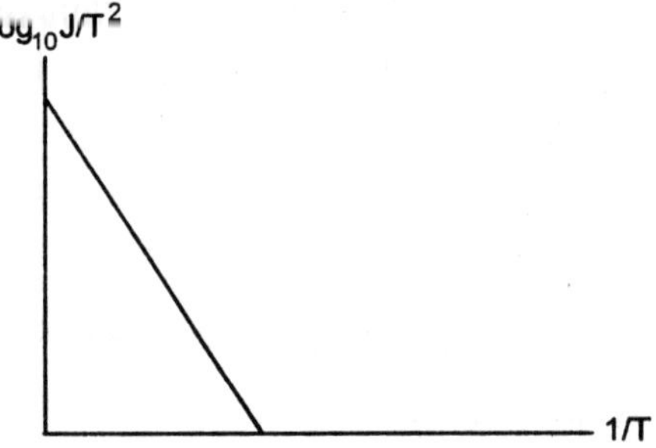

Fig. 6.3.2 Variation of $\log_{10} J/T^2$ with 1/T.

6.4 PROPERTIES OF IDEAL BOSE SYSTEM

In this section we shall discuss some peculiar properties of a Bose gas, which consists of bosons of non-zero mass.

It is customary to express the parameter α appearing in *B-E* statistics in terms of another parameter μ, called chemical potential, through relation

$$\alpha = -\beta\mu = -\frac{\mu}{kT} \tag{6.4.1}$$

For an ideal *B-E* gas of *N* molecules in an enclosure of volume *V*, the mean occupation number n_r in r-th single particle state (the most probable number of particles with energy ε_r)is

$$n_r(\varepsilon_r) = \frac{g_r}{\exp[\beta(\varepsilon_r - \mu)] - 1}, \ \beta = 1/kT$$

The parameter α or μ is determined as a function of *N* and temperature *T* by the condition

$$N = \sum_{r=0}^{\infty} n_r = \frac{g_1}{\exp[\beta(\varepsilon_1 - \mu)] - 1} + \frac{g_2}{\exp[\beta(\varepsilon_2 - \mu)] - 1} + \ldots\ldots.$$

$$= n_1 + n_2 + \ldots\ldots\ldots\ldots \tag{6.4.2}$$

where the sum is over the discrete energy levels. Replacing the summation by integration we have

$$N = \frac{2\pi V}{h^3}(2m)^{3/2} \int_0^{\infty} \frac{\varepsilon^{1/2} d\varepsilon}{e^{\beta(\varepsilon - \mu)} - 1} \tag{6.4.3}$$

The mean occupation number is always positive or zero *i.e.*, $n_r \geq 0$ for all values of ε. We take the energy scale such that the ground state energy ε_1 is zero. The occupation number of the ground state is

$$n_1 = \frac{g_1}{e^{-\beta\mu} - 1}, \ g_1 = 1, \varepsilon_1 = 0$$

The condition $n_1 \geq 0$ implies that $\mu = 0$. Thus μ of an ideal *B-E* gas is always negative. From Eq.(198) we see that left hand side *i.e.*, N is constant, so must be the right hand side. This implies that as T is lowered (or β is increased), α which is negative, must increase. Thus the maximum value of μ is zero, $\mu_{max} = 0$.

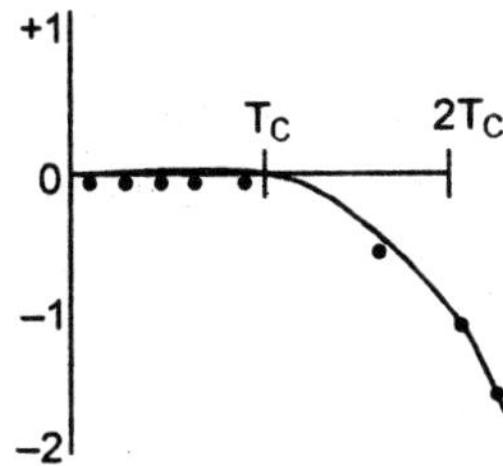

Fig. 6.4.1 Variation chemical potential μ with temperature T.

Let us define a minimum temperature T_C which corresponds to the maximum value of chemical potential *i.e.*, $\mu = 0$. This temperature is given by

$$N = \frac{2\pi V(2m)^{3/2}}{h^3} \int_0^\infty \frac{\varepsilon^{1/2}\, d\varepsilon}{e^{\varepsilon/kT_C} - 1} \tag{6.4.4}$$

To find the value of integral, we change the variable from ε to x by substitution $x = \varepsilon/kT_C$. Doing so we get

$$N = V\left(\frac{2\pi mkT_C}{h^2}\right)^{3/2} \left[\frac{2}{\sqrt{\pi}} \int_0^\infty \frac{x^{1/2}\, dx}{e^x - 1}\right]$$

The value of the quantity in square bracket is 2.61. So

$$N = V\left(\frac{2\pi mkT_C}{h^2}\right)^{3/2} (2.61) \tag{6.4.5}$$

$$N = \frac{V}{\lambda^3}(2.61) \tag{6.4.6}$$

where $\lambda = \dfrac{h}{\sqrt{2\pi mkT_C}}$ is the thermal de Broglie wavelength of molecule. The critical temperature T_C is given by

$$T_C = \frac{h^2}{2\pi mk}\left(\frac{\rho}{2.61}\right)^{2/3}, \qquad \rho = \frac{N}{V}.$$

For 1 mol of gas $N = N_A$

$T_C = \dfrac{115}{MV_m^{2/3}}$ kelvin, M = molar mass, V_m = molar volume in cm^3/mol.

The Eq.(6.4.3) has no solution for $T < T_C$. This problem arises due to replacement of discrete sum Σ by $\int$ in Eq.(6.4.3). This can be seen as follows.

We know that $g(\varepsilon) = \frac{2\pi V}{h^3}(2m)^{3/2}\,\varepsilon^{1/2}$. For ground state $\varepsilon_1 = 0$, $g(0) = 0$ and hence $n_1 = 0$. The occupation number corresponding to ground state energy does not contribute to the total number of particles. In fact $g(\varepsilon)$ for ground state is not zero but 1. At high temperature this replacement of Σ by $\int$ does not introduce any significant error because the ground state is thinly populated and the contribution of this term may be omitted. At very low temperature $T < T_C$ we cannot over look the first term in the sum (6.4.2). Instead we must explicitly retain the first term as such and write the remaining terms as integral as given below.

$$N = \frac{1}{e^{-\beta\mu} - 1} + \sum_{i=2}^{\infty} n_i = \frac{1}{e^{-\beta\mu} - 1} + \frac{2\pi V}{h^3}(2m)^{3/2}\int_0^\infty \frac{\varepsilon^{1/2}}{e^{\beta(\varepsilon-\mu)} - 1}\,d\varepsilon \qquad (6.4.7)$$

The first term in Eq.(6.4.7) represents the number of particles in the ground state ($\varepsilon_1 = 0$). These particles do not contribute to the energy and momentum of the system. The second term represents aggregate of particles occupying the higher energy state $\varepsilon > 0$. Only these particles contribute to the energy and momentum of the system.

At higher temperature ($T > T_C$) the number of particles in the ground state is negligibly small and hence may be omitted.

Below T_C, the chemical potential is very close to zero ($\mu \to 0$) and the number of particles in the non-zero energy states ($\varepsilon > 0$) is given by

$$N_{\varepsilon>0,\mu=0} = V\frac{2\pi}{h^3}(2m)^{3/2}\int_0^\infty \frac{\varepsilon^{1/2}\,d\varepsilon}{e^{\beta\varepsilon} - 1} \qquad (6.4.8)$$

$$= V\left(\frac{2\pi\, mkT}{h^2}\right)^{3/2}\left[\frac{2}{\sqrt{\pi}}\int_0^\infty \frac{x^{1/2}\,dx}{e^x - 1}\right], \qquad x = \beta\,\varepsilon \qquad (6.4.9)$$

$$= V\left(\frac{2\pi\, mkT}{h^2}\right)^{3/2}(2.612) \qquad (6.4.10)$$

From Eq.(6.4.5) and (6.4.10) we have

$$\frac{N_{\varepsilon>0}}{N} = \left(\frac{T}{T_C}\right)^{3/2} \qquad (6.4.11)$$

Eq.(6.4.11) gives the fraction of particles occupying the states with energy $\varepsilon > 0$. The fraction of particles occupying the ground state is given by

$$\frac{N_{\varepsilon=0}}{N} = 1 - \left(\frac{T}{T_c}\right)^{3/2} \qquad (6.4.12)$$

A plot of $N_{\varepsilon=0}/N$ as a function T/T_C is shown in the figure. From the figure, it is evident that for $T > T_C$, the number of particles in the ground state is negligible. As T decreases below T_C, the number of particles in the ground state increases rapidly. The process of dropping particles rapidly into the ground state with zero energy is known as *Bose-Einstein condensation.* The particles in this state possess zero energy and zero momentum. They contribute neither pressure nor do they possess viscosity. (Viscosity is related to transport of momentum.) B-E condensation is second order phase transition. A *B-E* gas at temperature below T_C is called *degenerate* and T_C is known as *degeneracy temperature* or *condensation temperature*.

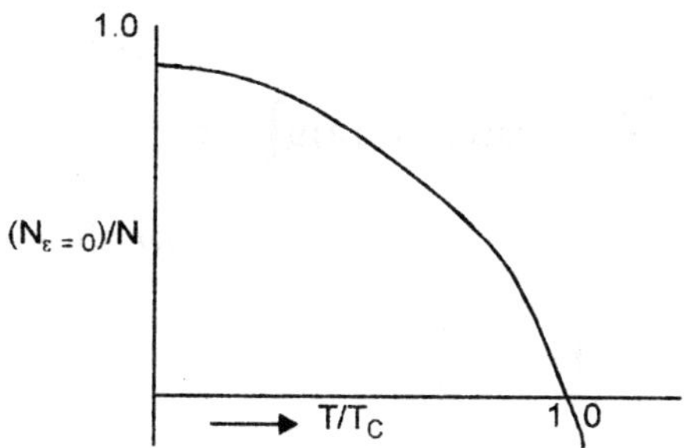

Fig. 6.4.2 Fraction of particles in the zero-energy ground state as function of temperature.

6.5 ENERGY OF *B-E* GAS

Below T_C, the particles occupying the ground state have no energy. Only the particles occupying the energy states $\varepsilon > 0$ have energy and the number of such particles is given by

$$N_{\varepsilon>0} = N\left(\frac{T}{T_C}\right)^{3/2}$$

The energy of each particle is of the order of kT. Hence the energy of B-E gas is

$$E = N_{\varepsilon>0}.kT = Nk\left(\frac{T^{5/2}}{T_C^{3/2}}\right) \quad \text{for } T < T_C. \tag{6.5.1}$$

The molar heat capacity is given by

$$C_V = \left(\frac{\partial E}{\partial T}\right)_V = \frac{5}{2}R\left(\frac{T}{T_C}\right)^{3/2} \quad \text{for } T < T_C \tag{6.5.2}$$

The exact calculation from the B-E distribution gives

$$C_V = 1.93R\left(\frac{T}{T_C}\right)^{3/2} \quad \text{for } T < T_C \tag{6.5.3}$$

The variation of C_V of a *B-E* gas with temperature is shown in the figure (6.5.3).

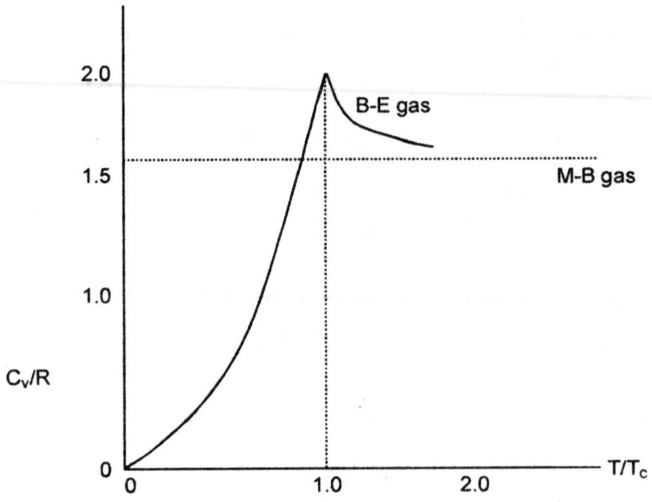

Fig. 6.5.3 Variation of C_V with temperature.

6.6 BLACK BODY RADIATION: PLANCK'S RADIATION LAW

The thermal radiation in a cavity maintained at temperature T is a well-known example of *B-E* system. In quantum picture the radiation is regarded as an assembly of particles, called photons, each of which has spin 1. The energy ε and momentum p of photon are given by $\varepsilon = \hbar\,\omega$ and $p = (\hbar\omega/c) = \varepsilon/c$.

To obtain the energy of photon gas we need to know the number of quantum states $g(p)\,dp$ available to photons with momentum in the range dp at p. Since photon has zero rest mass, the expression for density of states

$$g(\varepsilon)d\varepsilon = \frac{2\pi V}{h^3}(2m)^{3/2}\,\varepsilon^{1/2}\,d\varepsilon \tag{6.6.1}$$

as such can not be applied. To apply it to photons we must convert it in terms of momentum p through substitutions $\varepsilon = p^2/2m$ and $d\varepsilon = (2pdp/2m)$. Making use of this substitution we get

$$g(p)dp = \frac{4\pi V}{h^3}p^2 dp \tag{6.6.2}$$

This expression for the number of quantum states accessible to photons needs correction. Electromagnetic waves are purely transverse and there can be two sets of waves polarized in mutually perpendicular planes or right and left circular polarizations. Thus a photon of definite momentum can be in two possible states. The net effect is to multiply the above expression for the density of states by two. Thus the number of photon states in which the photon has momentum in the range p and $p + dp$ is

$$g(p)dp = \frac{8\pi V}{h^3}p^2 dp \tag{6.6.3}$$

In terms of frequency ω, the number of states in the frequency range $d\omega$ at ω is

$$g(\omega)d\omega = \frac{8\pi V}{h^3}\left(\frac{\hbar\omega}{c}\right)^2 d\left(\frac{\hbar\omega}{c}\right)$$

$$= \frac{V}{\pi^2 c^3} \omega^2 d\omega \qquad (6.6.4)$$

According to *B-E* distribution the mean number of photons per quantum state at energy ε is

$$f(\varepsilon) = \frac{1}{e^{\beta\varepsilon} - 1} = \frac{1}{e^{\beta\hbar\omega} - 1}, \qquad \beta = \frac{1}{kT} \qquad (6.6.5)$$

The number of photons in the frequency range ω and ω + *d*ω is

$$n(\omega)d\omega = f(\omega)g(\omega)d\omega = \frac{V}{\pi^2 c^3} \cdot \frac{\omega^2 d\omega}{e^{\beta\hbar\omega} - 1} \qquad (6.6.6)$$

The energy of photon gas in the frequency range ω and ω + *d*ω is

$$\varepsilon(\omega)d\omega = \hbar\omega . n(\omega)\, d\omega = \left(\frac{V\hbar}{\pi^2 c^3}\right)\left(\frac{\omega^3 d\omega}{e^{\beta\hbar\omega} - 1}\right) \qquad (6.6.7)$$

The energy density in the frequency range ω and ω + *d*ω is

$$u(\omega, T)d\omega = \frac{\hbar}{\pi^2 c^3} \cdot \frac{\omega^3}{e^{\beta\hbar\omega} - 1} d\omega \qquad (6.6.8)$$

or

$$u(\lambda, T)d\lambda = \frac{16\pi^2 \hbar c}{\lambda^5} \frac{1}{[\exp(2\pi\hbar c / \lambda kT)] - 1} d\lambda \qquad (6.6.9)$$

This is the *Planck's radiation law*. Stefan's law and Wien's law both have been derived from this law.

6.7 COMPARISON OF *M-B*, *B-E* AND *F-D* STATISTICS

(1) The distribution functions for the three statistics giving the mean number $\bar{n}_i$ of particles occupying a state with energy ε_i are given by

$$\bar{n}_i = g_i e^{-\alpha} e^{-\beta\varepsilon_i} \qquad (M\text{-}B)$$

$$\bar{n}_i = \frac{g_i}{e^{\alpha} e^{\beta\varepsilon_i} - 1} \qquad (B\text{-}E)$$

$$\bar{n}_i = \frac{g_i}{e^{\alpha} e^{\beta\varepsilon_i} + 1} \qquad (F\text{-}D)$$

In the classical limit $g_i >> n_i$ *B-E* and *F-D* both distribution functions approach the *M-B* distribution function. *M-B* statistics is a classical statistics, *B-E* and *F-D* statistics are quantum statistics.

(2) *M-B* statistics applies to systems comprising of *distinguishable* particles, whereas *B-E* and *F-D* statistics apply to *indistinguishable* particles.

(3) Spin of particles constituting the system in not a criterion for the applicability of M-B statistics. *B-E* statistics is applicable to particles having integral spins 0, 1, 2, ...Such particles are called *bosons*. Examples of *bosons* are: photons, phonons, hydrogen molecule, liquid helium, mesons, etc. *F-D* statistics applies to particles having half-integral spins

$\frac{1}{2}, \frac{3}{2}, \frac{5}{2}$ .....Such particles are called *fermions*. Examples of fermions are: electrons, protons, neutrons, ...etc.

(4) *M-B* and *B-E* statistics put no restriction on the number of particles that can occupy a quantum state. *F-D* statistics permits at the most one particle that a quantum state can accommodate.

(5) To specify a microstate of a system, the classical phase space is divided into cells whose volume may be taken as small as we please. A cell in this phase space represents a microstate of the system. In quantum mechanical description the phase space is divided into cells whose volume is not less than h^3, h being Planck's constant. A cell of volume h^3 represents a quantum state (microstate) of the system.

(6) At high temperature both quantum statistics (*B-E* and *F-D*) approach the *M-B* statistics.

(7) The variation of three distribution functions $f(\varepsilon) = n_i/g_i$, with energy at different temperatures is shown in the figures. According to *M-B* and *B-E* distribution functions, at a given temperature, particles like to occupy the lower enrgy states. At lower energy the occupation number is larger for *B-E* than for *M-B* distribution function. As the energy of the system increases, the occupation number decreases. According to *F-D* distribution function, at T = 0 K, all the quantum states with energy less than Fermi energy are occupied by electrons in an electron gas and those above the Fermi level are empty. As temperature increases, the electrons in the energy states a little below the Fermi level are excited to empty energy states a little above the Fermi level. At very high temperature the *F-D* distribution becomes more and more like *M-B* distribution.

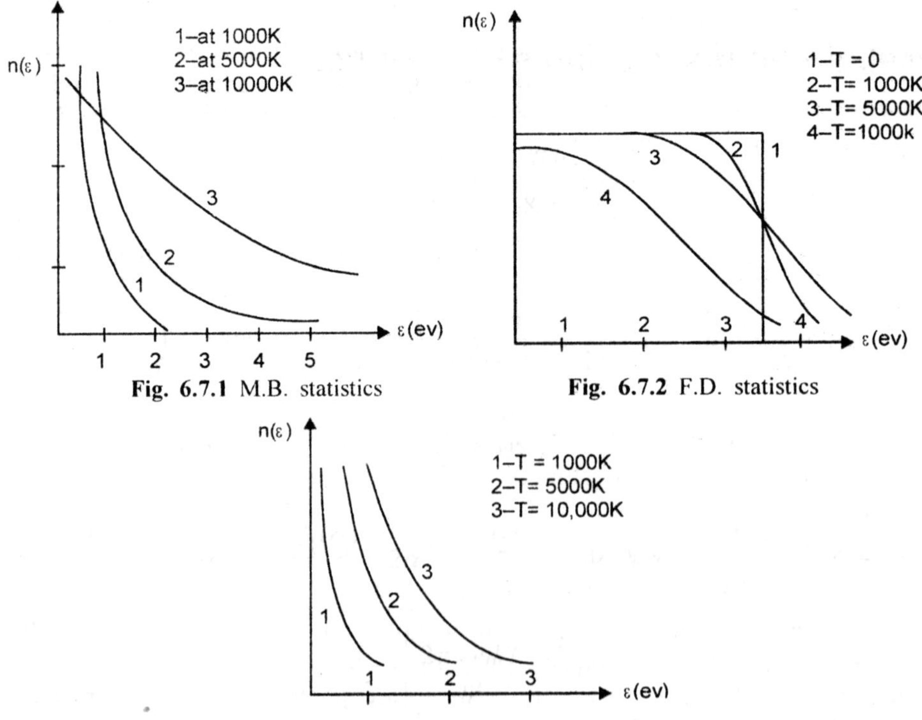

Fig. 6.7.1 M.B. statistics

Fig. 6.7.2 F.D. statistics

(8) The values of g_i/n_i for the three distribution functions are

$$M\text{-}B \qquad \frac{g_i}{n_i} = e^{\alpha+\beta\,\varepsilon_i}$$

$$B\text{-}E \qquad \frac{g_i}{n_i} = \frac{g_i}{e^{\alpha+\beta\,\varepsilon_i} - 1}$$

$$F\text{-}D \qquad \frac{g_i}{n_i} = \frac{g_i}{e^{\alpha+\beta\,\varepsilon_i} + 1}$$

When the number of quantum states in an energy level is much larger than the number of indistinguishable particles (bosons or fermions) *i.e.*, $g_i >> n_i$, the term 1 may be omitted from the denominator. In this situation, both the quantum statistics reduce to *M-B* statistics and hence *M-B* statistics can safely be used to the system. Thus the classical limit is reached when $g_i >> n_i$. This condition may be put in other forms as

$$e^{\alpha} >> 1$$

or
$$\left(\frac{V}{N}\right)\left(\frac{2\pi mkT}{h^2}\right)^{3/2} >> 1$$

or
$$a >> \lambda$$

where V = volume of the system, N = total number of particles in the system, T = temperature of the system, a = average separation of particles and λ = de Broglie wavelength of the particles.

6.8 VALIDITY CRITERION FOR CLASSICAL REGIME

The mean occupation number $n(\varepsilon)$ of energy state ε, according to classical (*M-B*) statistics, is

$$n(\varepsilon) = \frac{g_i}{e^{\alpha} e^{\varepsilon/kT}} = g_i e^{-\alpha} \quad e^{-MkT} \tag{6.8.1}$$

and that according to quantum statistics is

$$n(\varepsilon) = \frac{g_i}{e^{\alpha}\, e^{\varepsilon/kT} \pm 1} \tag{6.8.2}$$

+ sign for *F-D* and – sign for *B-E* statistics.

The quantum statistics become identical with the *M-B* statistics in the limit

$$e^{\alpha} \quad e^{MkT} \quad >> \quad 1 \tag{6.8.3}$$

for all values of ε. For $\varepsilon = 0$ the condition (6.8.3) reduces to

$$e^{\alpha} \quad >> \quad 1 \tag{6.8.4}$$

Condition (6.8.4)) will certainly hold for $\varepsilon > 0$. So the equation (6.8.4) represents the criterion for the validity of classical statistics. Now the parameter α is determined from the condition

$$N = \sum_i n_i = \sum_i g_i(\varepsilon_i) e^{-\alpha} \; e^{-\beta\varepsilon_i}$$

$$N\ e^{\alpha} = \sum_i g_i(\varepsilon_i)\ e^{-\beta\,\varepsilon_i} \Rightarrow \frac{2\pi\ V}{h^3}(2m)^{3/2}\int_0^{\infty}\varepsilon^{1/2}\ e^{-\beta\,\varepsilon}d\varepsilon$$

$$= \frac{2\pi\ V}{h^3}(2m)^{3/2}\ \frac{1}{2}\sqrt{\pi}$$

$$\therefore\ e^{\alpha} = \left(\frac{V}{N}\right)\left(\frac{2\pi\ mkT}{h^2}\right)^{3/2} \tag{6.8.5}$$

So, in view of Eq. (6.8.5) the criterion (6.8.4) becomes

$$\left(\frac{V}{N}\right)\left(\frac{2\pi\ mkT}{h^2}\right)^{3/2} >> 1 \tag{6.8.6}$$

If a is the average separation between the particles of the system then each particle may be allotted a cubical volume a^3. This must be equal to V/N. So $(V/N)^{1/3}$ gives the mean distance between the molecules. Now the thermal de Broglie wavelength of particles is given by

$$\lambda = \frac{h}{p} = \frac{h}{\sqrt{2m\varepsilon}} = \frac{h}{\sqrt{3mkT}},\quad \varepsilon = \frac{3}{2}kT$$

So,
$$\frac{1}{\lambda^3} = \left(\frac{3mkT}{h^2}\right)^{3/2} \approx \left(\frac{2\pi\ mkT}{h^2}\right)^{3/2} \tag{6.8.7}$$

In view of Eq.(6.8.7) the condition becomes

$$\left(\frac{a}{\lambda}\right)^3 >> 1$$

$$a >> \lambda \tag{6.8.8}$$

Thus the classical statistics is valid if the average separation between particles is much greater than the mean de Broglie wavelength of the particles. This condition will be satisfied when (*i*) the temperature is large (*ii*) number density is very small (*i.e.*, the gas is dilute) and (*iii*) mass of particle is not too small. When these conditions are not met, the particles are close together and their wave function overlap and they are no longer distinguishable.

Let us illustrate this by example. Consider helium gas at N.T.P. The average separation between molecules is

$$a = \left(\frac{V}{N}\right)^{1/3} = \left(\frac{22.4 \times 10^3}{6.6 \times 10^{23}}\right)^{1/3} = 32 \times 10^{-8}\,\text{cm} = 32\ \text{Å}$$

The de Broglie wavelength of molecule is

$$\lambda = \frac{h}{\sqrt{3mkT}} = \frac{6.6 \times 10^{-34}\,js}{\sqrt{3 \times 6.8 \times 10^{-27}\,\text{kg} \times 1.38 \times 10^{-23}\,J/K \times 300}} = 0.8 \times 10^{-10}\,\text{m} = 0.8\ \text{Å}$$

Since $a >> \lambda$, classical (*M-B*) statistics can be applied to helium gas at room temperature. Now consider liquid helium gas at 10 K. The average separation between molecules is

$$a = \left(\frac{V}{N}\right)^{1/3} = (5 \times 10^{-23} \ \text{cm}^3)^{1/3} \approx 4 \times 10^{-8}\,\text{cm} = 4\,\text{Å}$$

de Broglie wavelength of molecule is

$$\lambda = \frac{h}{\sqrt{3mkT}} = \frac{6.6 \times 10^{-34}\,Js}{\sqrt{3 \times 6.8 \times 10^{-27}\,\text{kg} \times 1.38 \times 10^{-23}\,J/K \times 10K}} = 4 \times 10^{-10}\,\text{m} = 4\,\text{Å}$$

Since $a \approx \lambda$, quantum (*B-E*) statistics must be applied to liquid helium.

For conduction electrons in metals the average separation between electrons is

$$a = \left(\frac{V}{N}\right)^{1/3} = \left(10^{-23} \ \text{cm}^3\right)^{1/3} \approx \ 2 \times 10^{-8}\,\text{cm} = 2\,\text{Å}$$

The de Broglie wavelength of electron is

$$\lambda = \frac{h}{\sqrt{3mkT}} = \frac{6.6 \times 10^{-34}\,Js}{\sqrt{3 \times 9.1 \times 10^{-31}\,\text{kg} \times 1.38 \times 10^{-23}\,J/K \times 300K}} = 62 \times 10^{-10}\,\text{m} = 62 \ \text{Å}$$

Since the condition $\lambda >> a$ is satisfied, quantum (*F-D*) statistics is most appropriate for the treatment of conduction electrons in metals.

7

Partition Function

7.1 CANONICAL PARTITION FUNCTION

The canonical partition function of a system of N-particles occupying a volume V at temperature T is defined by

$$Z(V,N,T)=\sum_r e^{-\varepsilon_r/kT} \tag{7.1.1}$$

where the summation is over all discrete quantum states accessible to the system. If the energies of these different states are not different we can write Z as follows

$$Z(N,V,T)=\sum_{\text{energy levels}} g(\varepsilon_r)\; e^{-\varepsilon_r/kT} \tag{7.1.2}$$

where summation is only over all different energies ε_r and $g(\varepsilon_r)$ is the number of quantum states possessing energy ε_r. The quantity Z is very useful for calculating the macroscopic properties of any system in equilibrium. The evaluation of the sum in Eq.(7.1.1) requires the knowledge of single particle quantum states of the constituent particles of the system.

There are many problems in which the Hamiltonian can be written as a sum of simpler Hamiltonians. The most obvious example is the case of a dilute monatomic gas where the molecules are on the average far apart and hence their intermolecular interaction can be neglected. The molecules of the gas have kinetic energies only. The total Hamiltonian is in this case is expressed as

$$H=\sum_{i=1}^{N}\frac{p_i^2}{2m} \tag{7.1.3}$$

N is the number of particles (molecules) in the gas.

Another example is the decomposition of Hamiltonian of a polyatomic molecule into its various degrees of freedom viz translational, rotational, vibrational, electronic etc.

$$H=H_{trans}+H_{rot}+H_{vib}+H_{ele} \tag{7.1.4}$$

There are many other problems in physics in which the Hamiltonian by a proper and clever choice of variables can be written as a sum of individual terms as shown above. In all such cases the partition function of a molecule comes out to be a product of partition functions corresponding to each degree of freedom. For example, consider a system comprising of particles a, b, c, ... Let the energy states of these particles be $\{\varepsilon_j^a\}$, $\{\varepsilon_k^b\}$, $\{\varepsilon_l^c\}$... The superscripts denote the particle and the subscript the energy state. In this case the partition function of the system becomes

$$Z(N,V,T) = \sum_{j,k,l...} e^{-[\varepsilon_j^a + \varepsilon_k^b + \varepsilon_l^c + ...]/kT}$$

$$= \sum_j e^{-\varepsilon_j^a/kT} \sum_k e^{-\varepsilon_k^b/kT} \sum_l e^{-\varepsilon_l^c/kT}$$

$$= Z_a . Z_b . Z_c ...... \quad (7.1.5)$$

This is a very important result. It shows that if we write the N-particle Hamiltonian as a sum of independent terms, then the calculation of Z reduces to a calculation of partition function z of a single particle. Since z requires knowledge only of energy levels of a single particle, its evaluation is quite simple.

7.2 CLASSICAL PARTITION FUNCTION OF A SYSTEM CONTAINING *N* DISTINGUISHABLE PARTICLES

In classical approximation, the energy of a N-particle system depends on $3N$ generalized coordinates q_1, ...q_{3N}, and $3N$ momentum coordinates p_1, ...p_{3N}. The phase space is divided into cells of volume h^{3N}. To evaluate the partition function from Eq. (7.1.1) we first take the sum over the number $\frac{dq_1....dq_{3N}.dp_1......dp_{3N}}{h^{3N}}$ of cells of phase space at point (q_1,....q_{3N}, p_1......p_{3N}) and then take the sum (or integral) over all such volume elements. Thus in classical approximation the partition function of a N-particle system is given by

$$Z_c = \int ..... \int e^{-\varepsilon(q,p)/kT} . \frac{dq_1.....dq_{3N}.dp_1......dp_{3N}}{h^{3N}} \quad (7.2.1)$$

Notice that the transition from quantum partition function to classical partition function can be accomplished by following replacement.

$$\sum_r \quad \rightarrow \quad \int \frac{dq_1.....dq_{3N}.dp_1......dp_{3N}}{h^{3N}} \quad (7.2.2)$$

For a system of particles possessing only translational kinetic energy, the evaluation of integral in Eq.(7.2.1) is simple.

$$Z = \frac{1}{h^{3N}} \int d^3q_1......d^3q_N \int \exp\left(-\frac{\beta(p_1^2 + p_2^2 + ...p_N^2)}{2m}\right) d^3p_1......d^3p_N$$

where $d^3q_1 = dxdydz$ and $d^3p_1 = dp_x dp_y dp_z$ etc.

$$Z = \frac{V^N}{h^{3N}} \int_{-\infty}^{\infty} \exp\left(-\frac{\beta\ p_1^2}{2m}\right) d^3p_1 .... \int_{-\infty}^{\infty} \exp\left(-\frac{\beta\ p_N^2}{2m}\right) d^3p_N$$

$$= \frac{V^N}{h^{3N}}\left(\frac{2\pi\ m}{\beta}\right)^{3/2} \text{..........}N \quad factors$$

$$= \frac{V^N}{h^{3N}}\left(\frac{2\pi\ m}{\beta}\right)^{3N/2}$$

$$= V^N\left(\frac{2\pi\ m}{\beta\ h^2}\right)^{3N/2} \tag{7.2.3}$$

We have already mentioned that the evaluation of partition function for many-particle system reduces to the evaluation of partition function for a single particle. Consider a N-particle system whose constituent particles have translational kinetic energy $\varepsilon_r = p^2/2m$ only. Let us illustrate this. In the phase space of a single particle, in the volume element $dxdydzdp_x dp_y dp_z$ ($= d^3q d^3p$) there are $\frac{dxdydzdp_x dp_y dp_z}{h^3}$ or $\frac{d^3qd^3p}{h^3}$ possible states. Therefore the partition function for a single particle is

$$z = \frac{1}{h^3}\int\text{......}\int\left\{\exp\left(-\frac{\beta p^2}{2m}\right)\right\}.d^3qd^3p$$

The integration over the ordinary space coordinates gives the volume V occupied by the system of particles. So

$$z = \frac{V}{h^3}\int_{-\infty}^{\infty}\int_{-\infty}^{\infty}\int_{-\infty}^{\infty}\left\{\exp\left(-\frac{\beta(p_x^2+p_y^2+p_z^2)}{2m}\right)\right\}dp_x dp_y dp_z$$

$$= \frac{V}{h^3}\int_{-\infty}^{\infty}\left\{\exp\left(-\frac{\beta p_x^2}{2m}\right)\right\}dp_x \int_{-\infty}^{\infty}\left\{\exp\left(-\frac{\beta p_y^2}{2m}\right)\right\}dp_y \int_{-\infty}^{\infty}\left\{\exp\left(-\frac{\beta p_z^2}{2m}\right)\right\}dp_z$$

$$= \frac{V}{h^3}\left(\frac{2\pi\ m}{\beta}\right)^{1/2}\left(\frac{2\pi\ m}{\beta}\right)^{1/2}\left(\frac{2\pi\ m}{\beta}\right)^{1/2}$$

$$= V\left(\frac{2\pi mkT}{h^2}\right)^{3/2}$$

The partition function for the indistinguishable N-particle system is

$$Z = z^N = V^N\left(\frac{2\pi\ mkT}{h^2}\right)^{3N/2} \tag{7.2.4}$$

$$\ln Z = N\left[\ln V + \frac{3}{2}\ln\left(\frac{2\pi\ m}{h^2}\right) - \frac{3}{2}\ln\beta\right], \qquad \beta = 1/kT \tag{7.2.5}$$

7.3 THERMODYNAMIC FUNCTIONS OF MONATOMIC GAS

Mean energy
$$\overline{E} = -\frac{\partial \ln Z}{\partial \beta} = \frac{3N}{2\beta} = \frac{3}{2} NkT \tag{7.3.1}$$

Helmholtz free energy
$$F = -kT \ln Z = -NkT \ln\left[\left(\frac{2\pi \ mkT}{h^2}\right)^{3/2} V\right] \tag{7.3.2}$$

The entropy of the system
$$S = k\left[\ln Z + \beta\,\overline{E}\right]$$
$$= Nk\left[\ln V + \frac{3}{2}\ln T + \frac{3}{2}\ln\left(\frac{2\pi\, mk}{h^2}\right) + \frac{3}{2}\right] \tag{7.3.3}$$
$$= Nk\left[\ln V + \frac{3}{2}\ln T + \sigma\right] \tag{7.3.4}$$

where
$$\sigma = \frac{3}{2}\left[\ln\left(\frac{2\pi\, mk}{h^2}\right) + 1\right] \tag{7.3.5}$$

The application of equation (7.3.4) in calculating the change in entropy when two samples of the same gases of equal volume and at the same pressure and temperature are mixed leads to the famous paradox known as **Gibbs paradox**. This equation is valid for a gas of *distinguishable* molecules and needs correction when it is to be applied to a system of *indistinguishable* molecules

7.4 GIBBS PARADOX

Consider a vessel divided into two compartments by a removable partition. The two compartments are then filled with two different gases. The number of molecules and volumes of the two gases are N_1, V_1 and N_2, V_2 as shown in the figure. The gases are at the same temperature and pressure. Now the partition is removed. On removing thr partition, the gases are mixed owing

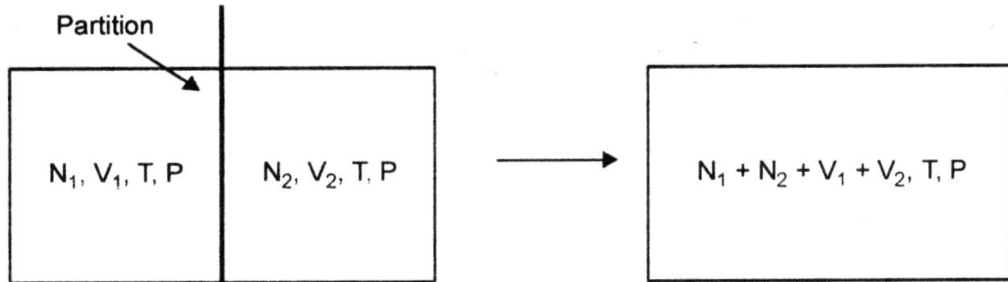

Fig. 7.4.1 Mixing of two different gases initially separated by a partition wall.

to diffusion of molecules. The diffusion is a irreversible process. (By putting the partition back in its original position, the gases don't unmix and the original state of gases is not achieved.) In irreversible process of mixing of the two different gases the entropy increases. Let us calculate the increase in entropy. The entropy of a gas composed of N identical distinguishable molecules is given by

$$S = Nk\left[\ln V + \frac{3}{2}\ln T + \frac{3}{2}\left(\ln\frac{2\pi\ mk}{h^2}\right) + \frac{3}{2}\right] \tag{7.4.1}$$

Before mixing the entropy of the combined system is

$$S_{12} = S_1 + S_2 = N_1 k\left[\ln V_1 + \frac{3}{2}\ln T + \frac{3}{2}\left(\ln\frac{2\pi\ m_1 k}{h^2}\right) + \frac{3}{2}\right] +$$

$$N_2 k\left[\ln V_2 + \frac{3}{2}\ln T + \frac{3}{2}\left(\ln\frac{2\pi\ m_2 k}{h^2}\right) + \frac{3}{2}\right] \tag{7.4.2}$$

After mixing the entropy of the combined system is

$$S'_{12} = S'_1 + S'_2 = N_1 k\left[\ln(V_1 + V_2) + \frac{3}{2}\ln T + \frac{3}{2}\left(\ln\frac{2\pi\ m_1 k}{h^2}\right) + \frac{3}{2}\right] +$$

$$N_2 k\left[\ln(V_1 + V_2) + \frac{3}{2}\ln T + \frac{3}{2}\left(\ln\frac{2\pi\ m_2 k}{h^2}\right) + \frac{3}{2}\right] \tag{7.4.3}$$

The change in entropy is

$$\Delta S = k\left[N_1 \ln\frac{V_1 + V_2}{V_1} + N_2 \ln\frac{V_1 + V_2}{V_2}\right] \tag{7.4.4}$$

If we take $N_1 = N_2 = N$ and $V_1 = V_2 = V$ then

$$\Delta S = 2\text{ k } N \ln 2 = \text{positive number.} \tag{7.4.5}$$

The entropy in the process increases and this result is in agreement with the experiments.

Now suppose that the two compartments of the vessel are filled with the same gases such that $N_1 = N_2 = N$, $V_1 + V_2 = V$. The two samples of the gases are at the same temperature and pressure. If the partition is removed, the increase in entropy of the combined system calculated as before comes out to be equal to

$$\Delta S = 2 N k \ln 2$$

That is, the entropy increases in the process of mixing of two identical samples of the same gases. This conclusion is not correct. Further if we put a large number of partitions in the vessel and remove them one by one, then by increasing the entropy in each act of removing the partition,

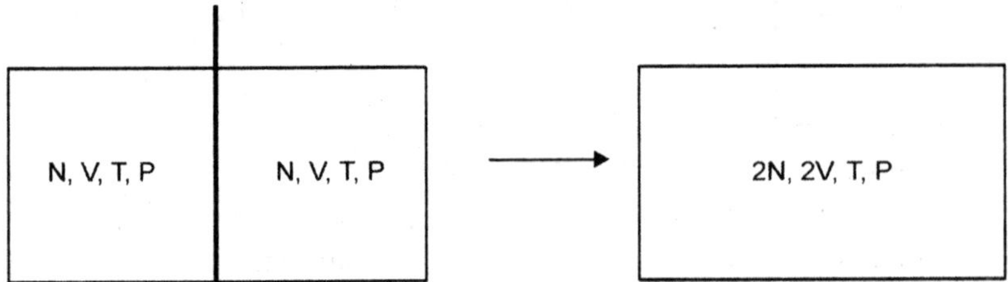

Fig. 7.4.6 Mixing of two identical samples of the same gases

we can increase the entropy of the combined system by any amount. The mixing of two identical samples of the same gases is a reversible process because by inserting the partition to its initial position, we get the state of the gases, which are in no way different from that we had before mixing. The total entropy of the system should not change on removal of the partition. This is known as Gibbs paradox. The origin of this paradox lies in the use of expression for entropy derived from the formula of partition function

$$Z = z^N = V^N \left(\frac{2\pi\ mkT}{h^2}\right)^{3N/2} \quad \textit{(distinguishable particles)} \qquad (7.4.6)$$

In the derivation of this expression it was assumed that the particles of the system are *distinguishable* and the interchange of positions of two molecules would lead to a physically distinct state of the system. But this is not so. In quantum mechanical treatment of a gas, the molecules are completely *indistinguishable*. A calculation of partition function and entropy, assuming the indistinguishability of molecues, would not give rise to Gibbs paradox. A way out to the Gibbs paradox is to apply a correction to the expression of partition function (7.4.6). We know that N molecules can be arranged in $N!$ ways by permuting among themselves. If the molecules are considered indistinguishable, then these $N!$ possible permutations of such molecules would not lead to physically distinct states. So the number of distinct states over which the summation is made in calculation of classical partition function is large by a factor $N!$. The correct partition function will be that which takes into account the indistinguishability of molecules. This is obtained by dividing the expression (7.4.6) of partition function by $N!$.

$$Z = \frac{z^N}{N\ !} = \frac{1}{N\ !}\left[V\left(\frac{2\pi\ m}{\beta\ h^2}\right)^{3/2}\right]^N \quad \textit{(indistinguishable particles)} \qquad (7.4.7)$$

$$\ln Z = N\left[\ln V + \frac{3}{2}\ln\left(\frac{2\pi\ m}{\beta\ h^2}\right)\right] - \ln N!$$

Using Stirling's approximation $\ln N! = N \ln N - N$, we obtain

$$\ln Z = N\left[\ln V + \frac{3}{2}\ln\left(\frac{2\pi\ m}{\beta\ h^2}\right)\right] - N\ln N + N$$

$$= N\left[\ln\frac{V}{N} + \frac{3}{2}\ln\left(\frac{2\pi\ m}{\beta\ h^2}\right) + 1\right] \qquad (7.4.8)$$

Entropy

$$S = k[\ln Z + \beta E], \qquad E = (3/2)\ N\ k\ T$$

Therefore

$$S = k\ln Z + \frac{3}{2}Nk$$

$$= Nk\left[\ln\frac{V}{N} + \frac{3}{2}\ln T + \frac{3}{2}\ln\left(\frac{2\pi\ mk}{h^2}\right) + \frac{5}{2}\right] \qquad (7.4.9)$$

Using this formula for calculating the increase in entropy in mixing of two samples of the same gases we obtain

$$S_{initial} = S_1 + S_2 = 2Nk\left[\ln\frac{V}{N} + \frac{3}{2}\ln T + \frac{3}{2}\ln\left(\frac{2\pi\ mk}{h^2}\right) + \frac{5}{2}\right] \tag{7.4.10}$$

$$S_{final} = S_1' + S_2' = 2Nk\left[\ln\frac{2V}{2N} + \frac{3}{2}\ln T + \frac{3}{2}\ln\left(\frac{2\pi\ mk}{h^2}\right) + \frac{5}{2}\right] \tag{7.4.11}$$

Change in entropy on mixing the gases

$$\Delta S = 0$$

7.5 INDISTINGUISHABILITY OF PARTICLES AND SYMMETRY OF WAVE FUNCTIONS

Let us consider a system consisting of two identical particles labeled 1 and 2. If the particle is restricted to move in a certain region, the quantum mechanical treatment of the particle allows it to have discrete quantum states and discrete energy levels. Let φ_r (1) and E_r (1) denote the *r*-th quantum state (wave function) and energy of particle 1. Similarly φ_s (2) and E_s (2) denote the *s*-th state and energy of particle 2. Now suppose that both the particles 1 and 2 are present simultaneously in the same region.

(*i*) If the average separation between the particles is much greater than their de Broglie wavelength *i.e.*, the wavefunctions of the particles donot overlap, the particles are said to be distinguishable and the wave function of the two particles is simple product of individual particles. Thus

$$\psi\ (1,2) = \phi_r\ (1)\ \phi_s\ (2) \tag{7.5.1}$$

ϕ_r (1) means that the particle 1 is in the state ϕ_r. This may be generalized to a system of N particles, where *N* is very large. The wavefunction of an *N*-particle system is

$$\psi\ (1,2,3.....) = \varphi_r\ (1)\ \varphi_s\ (2)\ \varphi_t\ (3)\ ....\ \varphi_z\ (N) \tag{7.5.2}$$

By distinguishable particles we mean that any interchange of particles among the occupied states viz φ_r (2) φ_s (1) leads to a new state for the system without any change in the total energy of the system. A distribution function, which gives the distribution of particles among the various energy levels, derived on the assumption that the particles are distinguishable, is called classical or Maxwell-Boltzmann distribution.

If the average separation between the particles is less than the de Broglie wavelengths of the particles, then their wave functions overlap. The particles are said to be indistinguishable and quantum statistics is appropriate for their description. The wave function of the whole system must satisfy certain symmetry requirements.

If the system is composed of particles having integral spin (0, 1, 2, ...), it must be described by a wave function that must be symmetric with respect to interchange of two particles. That is, the wave function should not change its sign on interchanging two particles.

$$\psi^s\ (1,2) = \psi^s\ (2,1)$$

The superscript *s* stands for symmetry. For a system of two particles, the wave function of the system is obtained from the linear combination of single particle wave functions φ_r (1) and φ_s (2). Thus

$$\psi^s(1,2) = \frac{1}{\sqrt{2}}\left[\varphi_r(1)\varphi_s(2) + \varphi_r(2)\varphi_s(1)\right] \tag{7.5.3}$$

where $\sqrt{2}$ is normalization factor. We can verify that the wave function (7.5.3) of the system remains unchanged on interchanging the particles.

$$\psi^s(2,1) = \frac{1}{\sqrt{2}}[\varphi_r(2)\varphi_s(1) + \varphi_r(1)\varphi_s(2)] = \psi^s(1,2)$$

Thus the wave function ψ^s (1,2) given by (7.5.3) satisfies the symmetry requirement. A system consisting of particles of integral spins are described by symmetric wave function and the statistical behaviour of the system is described by quantum statistics called Bose-Einstein statistics. The particles with integral spin are called *bosons*. If n_r represents number of bosons in any quantum state then n_r = 0, 1, 2, 3, ... There is no restriction on the number of bosons that can be in a quantum state.

A system consisting of particles having half-integral spins $(\frac{1}{2}, \frac{3}{2}, \ldots\ldots)$ must be described by wave function that must be anti-symmetric with respect to interchange of two particles. That is, the wave function must change sign without a change in its magnitude. For a two-particle wave function this requirement can be expressed as

$$\psi^A(1,2) = -\psi^A(2,1) \tag{7.5.4}$$

For a two-particle system, the wave function $\psi^A(1,2)$, which is anti-symmetric is obtained from the linear combination of single particle wave functions as follows.

$$\psi^A(1,2) = \frac{1}{\sqrt{2}}[\varphi_r(1)\varphi_s(2) - \varphi_r(2)\varphi_s(1)] \tag{7.5.5}$$

$$= \frac{1}{\sqrt{2}}\begin{vmatrix} \varphi_r(1) & \varphi_r(2) \\ \varphi_s(1) & \varphi_s(2) \end{vmatrix}$$

$$= -\psi^A(2,1)$$

For an N-particle system, the wave function of the system is given by Slater determinant

$$\psi^A(1,2,3,....N) = \frac{1}{\sqrt{N}}\begin{vmatrix} \varphi_r(1) & \varphi_r(2) & .... & \varphi_r(N) \\ \varphi_s(1) & \varphi_s(2) & ..... & \varphi_s(N) \\ & & & \\ \varphi_z(1) & \varphi_z(2) & ..... & \varphi_z(N) \end{vmatrix} \tag{7.5.6}$$

where $\sqrt{N}$ is normalization factor. The particles with half-integral spin are called *fermions*.

Putting $r = s$ in ψ^A (1, 2) given by (7.5.5) we have ψ^A (1, 2) = 0. That is if we put *fermions* in the same state then the wave function vanishes. In other words *no two fermions can be in the same quantum state.* This statement is called Pauli exclusion principle. If n_r is the number of fermions in any quantum state then n_r = 0, 1 for all r.

7.6 PARTITION FUNCTION FOR INDISTINGUISHABLE PARTICLES

Let the wave functions and energies of two non-interacting distinguishable particles 1 and 2 be ϕ_r (1), ε_r (1) and ϕ_s (2), ε_s (2) respectively. If both the particles are present simultaneously in a region, the combined wave function and energy of the system are given by

$$\psi\ (1,2) = \varphi_r\ (1)\ \varphi_s\ (2)$$
$$\varepsilon_{12} = \varepsilon_r\ (1) + \varepsilon_s\ (2)$$

The canonical partition function of the system is

$$Z = \sum_{r,s} e^{-\beta\varepsilon_{12}} = \sum_{r,s} e^{-\beta(\varepsilon_r(1)+\varepsilon_s(2))}$$

$$= \sum_r e^{-\beta\varepsilon_r(1)} \sum_s e^{-\beta\varepsilon_s(2)} = z_1 . z_2$$

where summation extends over all quantum states of the individual particles 1 and 2. If the particles are identical $\varepsilon_r\ (1) = \varepsilon_s\ (2)$ and $z_1 = z_2 = z$ (say). The partition function of the system becomes

$$Z = z^2$$

Generalization of this result to N-particle system gives

$$Z = z^N \qquad \text{(identical distinguishable particles)}$$

If the particles 1 and 2 are indistinguishable the wave function $\psi\ (1,2)$ of the system must be either symmetric or anti-symmetric. The total energy ε_{12} of the system can be expressed in $2! = 2$ ways as

$$\varepsilon_{12} = \varepsilon_r\ (1) + \varepsilon_s\ (2) \quad \text{or} \quad \varepsilon_{12} = \varepsilon_r\ (2) + \varepsilon_s\ (1)$$

These two ways of writing the total energy corresponds to a single wave function $\psi^s\ (1,2)$ or $\psi^A\ (1,2)$. The expression for partition function $Z = \sum_{r,s} e^{-\beta\varepsilon_{12}}$ for the system, contains two terms

$$\varepsilon_{12} = \varepsilon_r\ (1) + \varepsilon_s\ (2) \text{ and } \varepsilon_{12} = \varepsilon_r\ (2) + \varepsilon_s\ (1)$$

corresponding to the same energy. Actually Z should contain only one term. To obtain the correct expression for partition function for a system containing indistinguishable particles, Z should be divided by 2!.

$$Z = \frac{z^2}{2!}$$

This result may be generalized to a system of N indistinguishable particles. Thus

$$Z = \frac{z^N}{N!} \tag{7.6.1}$$

In the classical limit, when the number of states is much greater than the number of particles available *i.e.*, $g_i >> n_i$, the difference between bosons and fermions may be neglected and Maxwell-Boltzmann statistics along with the expression for partition function $Z = \dfrac{z^N}{N!}$ may be used without any appreciable error.

Partition Function Of a System Consisting Of *N*-Indistinguishable Particles

If the energy levels of all the particles are the same, then the partition function of a system of N identical, indistinguishable particles, satisfying the condition that the number of available quantum states is much greater then the number of particles, is

$$Z(N,V,T)=\frac{1}{N!}\left[\sum_r e^{-\varepsilon_r/kT}\right]^N=\frac{z^N}{N!} \tag{7.6.2}$$

The presence of factor $1/N!$ is in accordance with the rule of correct *Boltzmann counting*. Eq. (7.6.1) is an extremely important result since it reduces a many body problem to a one-body problem.

The partition function for the indistinguishable N-particle system is

$$Z(T,V,N=\frac{z^N}{N!}=\frac{V^N}{N!}\left(\frac{2\pi\ mkT}{h^2}\right)^{3N/2} \tag{7.6.3}$$

7.7 MOLECULAR PARTITION FUNCTION

To a first approximation, the internal degrees of freedom (vibrational, rotational, electronic, nuclear) may be assumed to be independent of each other and the total energy E may be expressed as the sum of translational, rotational, vibrational, electronic and nuclear energies.

$$E = E_t + E_r + E_v + E_e + E_n$$

The partition function of molecule is

$$z=\sum_i e^{-\beta E_i}=\sum_i e^{-\beta(E_t+E_r+E_v+E_e+E_n)}$$

$$=\left(\sum_i e^{-\beta E_t}\right)\left(\sum_i e^{-\beta E_r}\right)\left(\sum_i e^{-\beta E_v}\right)\left(\sum_i e^{-\beta E_e}\right)\left(\sum_i e^{-\beta E_n}\right)$$

$$= z_t.\ z_r\ .z_v\ .z_e\ .z_n \tag{7.7.1}$$

Thus the partition function of molecule is the product of the translational, rotational, vibrational, electronic and nuclear partition functions.

The partition function of a gas of N molecules is

$$Z=\frac{1}{N!}\left[z_t^N\ .z_r^N\ .z_v^N\ .z_e^N\ .z_n^N\right] \tag{7.7.2}$$

7.8 PARTITION FUNCTION AND THERMODYNAMIC PROPERTIES OF MONATOMIC IDEAL GAS

Consider a monatomic gas dilute enough so that intermolecular interactions can be neglected. This condition is achieved at pressure below 1 atmosphere and at temperature greater than room temperature. The number of available quantum states far exceeds the number of particles of the gas. Under this condition the molecules of the gas have only translational kinetic energy viz $\varepsilon = p^2/2m$. The expression for the partition function for a single molecule has been obtained earlier as

$$z=V\left(\frac{2\pi\ mkT}{h^2}\right)^{3/2}$$

The partition function for the entire gas is

$$Z = \frac{z^N}{N!\ h^{3N}} = \frac{V^N}{N!}\left(\frac{2\pi\ mkT}{h^2}\right)^{3N/2} \tag{7.8.1}$$

The same result can also be obtained as follows.

The translational energy states of a molecule confined to move in a cube of length L are given by

$$E_{n_x,n_y,n_z} = \frac{h^2}{8mL^2}(n_x^2 + n_y^2 + n_z^2) \tag{7.8.2}$$

The translational partition function of a molecule is

$$z = \sum_{n_x,n_y,n_z=1}^{\infty} \exp\left(-\beta\,\varepsilon_{n_x,n_y,n_z}\right)$$

$$= \sum_{n_x=1}^{\infty} \exp\left(-\frac{\beta h^2}{8mL^2}n_x^2\right) \sum_{n_y=1}^{\infty} \exp\left(-\frac{\beta h^2}{8mL^2}n_y^2\right) \sum_{n_z=1}^{\infty} \exp\left(-\frac{\beta h^2}{8mL^2}n_z^2\right)$$

$$= \left[\sum_{n=1}^{\infty} \exp\left(-\frac{\beta h^2}{8mL^2}n^2\right)\right]^3 \tag{7.8.3}$$

The expression on the right hand side of Eq. (7.8.3) can not be expressed in terms of any simple analytic function. So we shall evaluate it in classical approximation. For a microscopic particle the successive terms in the expression for energy differ very little and therefore the summation Σ in Eq.(7.8.3) can be replaced by integral. So

$$z = \left[\int_0^{\infty} \exp\left(-\frac{\beta h^2}{8mL^2}dn\right)\right]^3 \tag{7.8.4}$$

Making use of the standard results

$$\int_0^{\infty} x^n.\exp(-\alpha\ x^2).dx = \frac{1.3.......(n-1)}{(2\alpha)^{n/2}}.\frac{1}{2}\left(\frac{\pi}{\alpha}\right)^{1/2},\quad n = 2,4,6...$$

$$= \frac{2.4......(n-1)}{(2\alpha)^{(n+1)/2}},\quad n = 3,5,7....$$

$$I_0(\alpha) = \frac{1}{2}\sqrt{\frac{\pi}{\alpha}},\qquad I_1(\alpha) = \frac{1}{2\alpha},\qquad I_2(\alpha) = \frac{1}{4\alpha}\sqrt{\frac{\pi}{\alpha}},\qquad I_3(\alpha) = \frac{1}{2\alpha^2}$$

we get

$$z = \left[\frac{2\pi\ mkT}{h^2}\right]^{3/2}.V \tag{7.8.5}$$

The partition function for the gas is

$$Z = \frac{z^N}{N!} = \frac{V^N}{N!}\left[\frac{2\pi\ mkT}{h^2}\right]^{3N/2} \tag{7.8.6}$$

Aliter: We can still arrive at the same result via another method. The partition function for a single molecule is

$$z = \int_0^\infty g(\varepsilon)\ e^{-\beta\varepsilon} d\varepsilon\ =\ \frac{2\pi\ V}{h^3}(2m)^{3/2}\int_0^\infty \varepsilon^{1/2}\ e^{-\beta\varepsilon}\, d\varepsilon$$

Making the substitution $x = \beta\varepsilon$ we have

$$z = \frac{2\pi\ V}{h^3}\left(\frac{2m}{\beta}\right)^{3/2}\int_0^\infty x^{1/2}\ e^{-x}dx$$

$$= \frac{2\pi\ V}{h^3}\left(\frac{2m}{\beta}\right)^{3/2}\ \Gamma\ (3/2) \qquad \Gamma 3/2 = \tfrac{1}{2}\sqrt{\pi}.$$

$$= \frac{V}{h^3}\left(\frac{2\pi\ m}{\beta}\right)^{3/2} = V\left(\frac{2\pi\ mkT}{h^2}\right)^{3/2}$$

7.9 HELMHOLTZ FREE ENERGY *F*

In terms of partition function, Helmholtz free energy is given by

$$F\ =\ -kT\ \ln\ Z \tag{7.9.1}$$

Making use of Stirling approximation $N! = (N/e)^N$ in Eq.(7.8.6) we have

$$\ln\ Z = N\ \ln\left[\frac{eV}{N}\left(\frac{2\pi\ mkT}{h^2}\right)^{3/2}\right] \tag{7.9.2}$$

Therefore

$$F = -NkT\ \ln\left[\frac{eV}{N}\left(\frac{2\pi\ mkT}{h^2}\right)^{3/2}\right] \tag{7.9.2}$$

Translational energy of the system is

$$E = -\frac{\partial}{\partial\beta}\ln Z = \frac{\partial}{\partial\beta}\left[N\ln\left\{\frac{eV}{N}\left(\frac{2\pi\ m}{\beta\ h^2}\right)^{3/2}\right\}\right] = \frac{3N}{2\beta} = \frac{3}{2}NkT \tag{7.9.3}$$

The pressure of the gas is

$$P = -\left(\frac{\partial F}{\partial V}\right)_{T,N} = NkT\frac{\partial}{\partial V}\left[\ln\left\{\frac{eV}{N}\left(\frac{2\pi\, mkT}{h^2}\right)^{3/2}\right\}\right]_{T,N} = \frac{NkT}{V} \tag{7.9.4}$$

This gives the equation of state

$$PV\ =\ NkT$$

Entropy of the gas is given by

$$S = -\left(\frac{\partial F}{\partial T}\right)_{V,N} = Nk\,\frac{\partial}{\partial T}\left[T\,\ln\left\{\frac{eV}{N}\left(\frac{2\pi\, mkT}{h^2}\right)^{3/2}\right\}\right]_{V,N}$$

$$= Nk\left[\ln\frac{V}{N}+\frac{3}{2}\ln T+\frac{3}{2}\ln\frac{2\pi mk}{h^2}+\frac{5}{2}\right] \qquad (7.9.5)$$

Equation (7.9.5) is known as Sackur-Tetrode equation. This equation shows that the entropy S tends to infinity as T tends to zero. This clearly violates the third law of thermodynamics. Had we used the original definition of partition function viz $Z = \Sigma \exp(-\beta\varepsilon_r)$ this difficulty would not have crop up. The problem stems from the replacement of the sum by integral. In fact this replacement is not justified near $T = 0$ where the contribution to energy from the ground state is significant. Whereas in the evaluation of integral for partition function we take $\varepsilon = 0$ or $p = 0$ in the ground state. At higher temperature the contribution from ground state is insignificant and so the replacement of sum by integral causes no appreciable error.

7.10 ROTATIONAL PARTITION FUNCTION

A rigid diatomic molecule in which two atoms of masses m_1 and m_2 are separated by a fixed distance r_0 is an example of rigid rotator. The moment of inertia of molecule is

$$I = \frac{m_1 m_2}{m_1+m_2} r_0^2 = \mu\ r_0^2$$

The rotational energy levels of the molecule are

$$E_r = \frac{\hbar^2}{2I} J(J+1) \qquad (7.10.1)$$

where J = 0, 1, 2 .... are rotational quantum numbers. The J-th energy level is $(2J + 1)$ fold degenerate. Hence the partition function of the molecule is

$$z = \sum_{J=0}^{\infty}(2J+1)\ e^{-\left[\frac{\beta\hbar^2}{2I}J(J+1)\right]} \qquad (7.10.2)$$

$$= \sum_{J=0}^{\infty}(2J+1)\quad e^{-\frac{\Theta}{T}J(J+1)} \qquad (7.10.3)$$

where $\Theta = \dfrac{\hbar^2}{2kI}$ is rotational characteristic temperature.

(*i*) When I or T is very small, we can express the partition function as

$$z = 1 + 3\ e^{-2\Theta/T} + 5\ e^{-6\Theta/T} + \ldots\ldots smaller\quad terms \qquad (7.10.4)$$

At low temperature the thermal energy kT is not enough to take the molecules to higher energy levels. The lower rotational energy levels will be heavily populated.

(*ii*) At higher temperature, the rotational energy levels form continuum and hence E_r may be assumed to be continuous. The summation sign in (7.10.2) or (7.10.3) may be replaced by integral sign. Putting J(J+1) = x we can write (7.10.2) as

$$z = \int_0^\infty (2J+1)\ e^{-\frac{\beta \hbar^2}{2I}J(J+1)} dJ = \int_0^\infty e^{-\frac{\beta \hbar^2}{2I}x} dx$$

$$= \frac{2I}{\beta\ \hbar^2} = \frac{2I\ kT}{\hbar^2} = \frac{T}{\Theta} \quad (7.10.5)$$

The general expression for z taking symmetry into consideration comes out to be

$$z = \frac{2I\ kT}{\sigma\ \hbar^2} \quad (7.10.6)$$

where σ = 1 for heteronuclear (asymmetric) molecule and σ = 2 for symmetric linear molecule.

At higher temperature $T > \Theta$, for an ideal diatomic gas

$$Z = \left[z^N\right] = \left[\frac{T}{\sigma\ \Theta}\right]^N \quad (7.10.7)$$

Helmholtz free energy $$F = -NkT \ln \frac{T}{\sigma\ \Theta} \quad (7.10.8)$$

Mean energy $$\bar{E} = kT^2 \frac{\partial \ln Z}{\partial T} = NkT \quad (7.10.9)$$

Heat capacity $$C_v = \left(\frac{\partial E}{\partial T}\right)_V = Nk \quad (7.10.10)$$

Entropy $$S = k \ln Z + \frac{E}{T}$$

$$= Nk\left[\ln \frac{I\ T}{\sigma} + \ln \frac{2k}{\hbar^2} + 1\right] \quad (7.10.11)$$

7.11 VIBRATIONAL PARTITION FUNCTION

A diatomic molecule made up of two atoms of masses m_1 and m_2 joined by '*spring*' of force constant k, acts like a harmonic oscillator. The classical frequency ω of the oscillator is $\omega = \sqrt{\frac{k}{\mu}}$, where m is the reduced mass of the molecule. The vibrational energy levels of the molecule are given by

$$\varepsilon_n = \left(n + \frac{1}{2}\right)\hbar\omega, \quad n = 0,\ 1,\ 2,\ \ldots \quad (7.11.1)$$

The energy $\varepsilon_0 = \frac{1}{2}\ \hbar\omega$ is called the zero-point energy. The vibrational partition function of the molecule is

$$z = \sum_{n=0}^{\infty} e^{-\beta\hbar\omega\left(n+\frac{1}{2}\right)} = e^{-\frac{1}{2}\beta\hbar\omega} \sum_{n=0}^{\infty} e^{-\beta\hbar\omega n} \tag{7.11.2}$$

The expression $\sum_{n=0}^{\infty} e^{-\beta\hbar\omega n}$ can be written in an alternative form as

$$\sum_{n=0}^{\infty} e^{-\beta\hbar\omega n} = 1 + e^{-\beta\hbar\omega} + e^{-2\beta\hbar\omega} + ... = \frac{1}{1-e^{-\beta\hbar\omega}}$$

Therefore

$$z = e^{-\frac{1}{2}\beta\hbar\omega} \cdot \frac{1}{1-e^{-\beta\hbar\omega}} \tag{7.11.3}$$

$$= e^{-\Theta/2T} \frac{1}{1-e^{-\Theta/T}} \tag{7.11.4}$$

where $\Theta = \dfrac{\hbar\omega}{k}$ is vibrational characteristic temperature. For a system consisting of N distinguishable particles the partition function is

$$Z = z^N = e^{-\frac{1}{2}N\beta\hbar\omega}\left(\frac{1}{1-e^{-\beta\hbar\omega}}\right)^N$$

$$= e^{-N\Theta/2T}\left(\frac{1}{1-e^{-\Theta/T}}\right)^N \tag{7.11.5}$$

and

$$\ln Z = -\frac{1}{2}N\beta\hbar\omega - N\ln\left(1-e^{-\beta\hbar\omega}\right) \tag{7.11.6}$$

Helmholtz function

$$F = -kT\ln Z = \frac{1}{2}N\beta\hbar\omega + NkT\ln\left(1-e^{-\beta\hbar\omega}\right) \tag{7.11.7}$$

Mean energy of the system

$$E = -\frac{\partial \ln Z}{\partial \beta}$$

$$= N\hbar\omega\left[\frac{1}{2} + \frac{1}{e^{\beta\hbar\omega} - 1}\right] \tag{7.11.8}$$

Heat capacity

$$C_V = \left(\frac{\partial E}{\partial T}\right)_V = Nk\left(\frac{\hbar\omega}{kT}\right)^2 \frac{e^{\beta\hbar\omega}}{\left(e^{\beta\hbar\omega}-1\right)^2} \tag{7.11.9}$$

$$= Nk\left(\frac{\Theta}{T}\right)^2 \frac{e^{\Theta/T}}{\left(e^{\Theta/T}-1\right)^2} \tag{7.11.10}$$

Entropy $$S = k\ln Z + \frac{E}{T}$$

$$= Nk\ln\left(\frac{1}{1-e^{-\beta\,\hbar\omega}}\right) + Nk\left(\frac{\hbar\omega}{kT}\right)\left(\frac{1}{1-e^{-\beta\,\hbar\omega}}\right) + \frac{N\hbar\omega}{T} \tag{7.11.11}$$

(*i*) At high temperature $\beta\hbar\omega = \frac{\hbar\omega}{kT} << 1$, thermal energy kT is much greater than the spacing of energy levels. In this case the energy of the system is

$$E = N\hbar\omega\left[\frac{1}{2} + \frac{1}{(1+\beta\;\hbar\omega+.....)-1}\right]$$

$$= N\hbar\omega\left[\frac{1}{2}+\frac{1}{\beta\;\hbar\omega}\right]$$

$$= \frac{N}{\beta} = NkT$$

Therefore C_V = Nk.

(*ii*) At low temperature $\beta\hbar\omega = \frac{\hbar\omega}{kT} >> 1$. The energy of the system is

$$E = N\hbar\omega\left[\frac{1}{2}+e^{-\beta\,\hbar\omega}\right] \Rightarrow \frac{1}{2}N\hbar\omega$$

Almost all the molecules will be in the ground state and the energy is temperature independent. Therefore C_V will te ' to zero.

7.12 GRAND CANONICAL ENSEMBLE AND GRAND PARTITION FUNCTION

In a grand canonical ensemble, each system is enclosed in a container whose walls are both heat conducting and permeable to the passage of particles (molecules). The number of molecules in a system, therefore, can range over all possible values *i.e.*, each system is open with respect to the transport of matter. We construct a grand canonical ensemble by placing a collection of such systems in a large heat bath at temperature T and a large reservoir of molecules. After equilibrium is reached, the entire ensemble is isolated from its surroundings. Since the entire ensemble is at equilibrium with respect to the transport of heat and matter, each system is specified by volume V, temperature T and chemical potential μ.

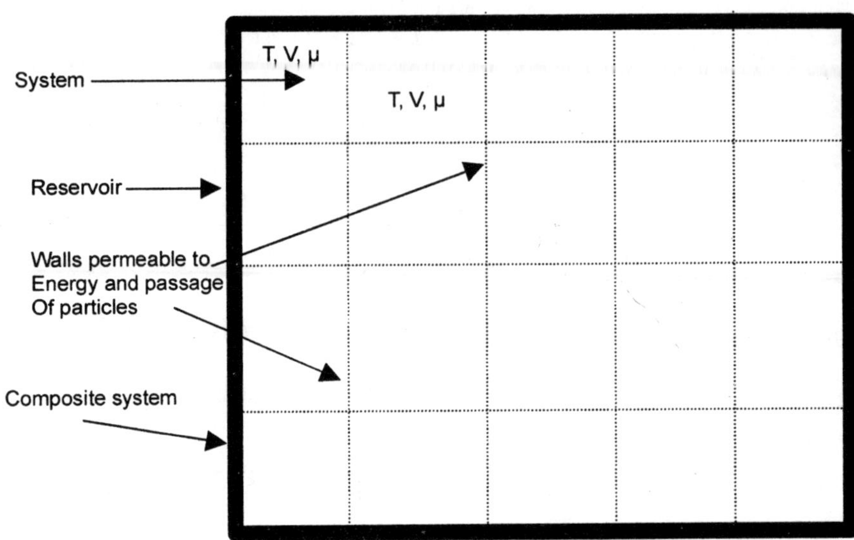

Fig. 7.12.1 Grand Canonical Ensemble. Each system has fixed volume V and temperature T but is open with respect to molecular transport.

Let the system A under study be in contact with a heat reservoir A'. The system A and A' constitute a composite system A^*. Let the system A be in its energy level E_r and has number of particles N_r. The corresponding values for heat reservoir are E' and N' and that of composite system are E^* and N^*. Since the composite system is insulated with respect to energy and passage of particles we must have

$$N_r + N' = N^* \tag{7.12.1}$$

$$E_r + E' = E^* \tag{7.12.2}$$

The probability $P_{N,r}$ that the system A in the ensemble is in the state r and contains N particles is proportional to the number Ω^* of states accessible to the composite system A*, which is just equal to the number $\Omega'(E', N')$ or $\Omega'(E^* - E_r, N^* - N_r)$ of states accessible to the reservoir. Thus

$$P_{N,r}(E_r, N_r) \propto \Omega'(E', N')$$

or

$$P_{N,r}(E_r, N_r) \propto \Omega'(E^* - E_r, N^* - N_r) \tag{7.12.3}$$

Since A is very small compared to A', $E_r << E^*$ and $N_r << N^*$, we can expand $\ln \Omega'$ in Taylor series as follows.

$$\ln \Omega'(E^* - E_r, N^* - N_r) = \ln \Omega'(E^*, N^*) - \left(\frac{\partial \ln \Omega'}{\partial E'}\right)_{E'=E^*} E_r - \left(\frac{\partial \ln \Omega'}{\partial N'}\right)_{N'=N^*} N_r$$

Denoting

$$\beta = \left(\frac{\partial \ln \Omega'}{\partial E'}\right)_{E'=E^*} \quad \text{and} \quad -\beta\ \mu = \left(\frac{\partial \ln \Omega'}{\partial N'}\right)_{N'=N^*} \tag{7.12.4}$$

we can write

$$\ln\Omega'(E^* - E_r, N^* - N_r) = \ln\Omega'(E^*, N^*) - \beta E_r + \beta\,\mu N_r$$

$$\Omega'(E^* - E_r, N^* - N_r) = \Omega'(E^*, N^*)\; e^{\beta(\mu N_r - E_r)} \tag{7.12.5}$$

The probability that any randomly chosen system contains N particles and be in r-th energy level with energy E_r is

$$P_{N,r} = C\,e^{\beta(\mu N_r - E_r)} \tag{7.12.6}$$

where C is a constant. Using the condition for normalization of probability viz. $\Sigma\, P_{Nr} = 1$ we have

$$C = \frac{1}{\sum\limits_{N,r} e^{\beta(\mu N_r - E_r)}} \tag{7.12.7}$$

Therefore,

$$P_{N,r} = \frac{e^{\beta(\mu N_r - E_r)}}{\sum\limits_{N,r} e^{\beta(\mu N_r - E_r)}} \tag{7.12.8}$$

The quantity in the denominator of right hand side of Eq.(7.12.7) is called *Grand Partition Function and denoted by* Z_G. Thus the grand partition function is

$$Z_G = \sum_{N=0}^{\infty}\sum_{r=0}^{\infty} e^{\beta(\mu N_r - E_r)} \tag{7.12.9}$$

$$= \left(\sum_{r=0}^{\infty} e^{-\beta E_r}\right) \sum_{N=0}^{\infty} e^{\beta\mu N_r}$$

$$= Z \sum_{N=0}^{\infty} e^{\beta\mu N_r} \tag{8.12.10}$$

Let $<n_{N,r}>$ denote the number of systems in the ensemble that contains N_r (variable) particles and are in the state r. Then

$$P_{N,r} = \frac{\langle n_{N,r}\rangle}{M} = \frac{e^{\beta(\mu N_r - E_r)}}{Z_G} \tag{7.12.11}$$

where M is total number of systems in the ensemble. The most probable distribution is then given by

$$n_{N,r} = \frac{M}{Z_G}\, e^{\beta(\mu N_r - E_r)} \tag{7.12.12}$$

7.13 STATISTICAL PROPERTIES OF A THERMODYNAMIC SYSTEM IN TERMS OF GRAND PARTITION FUNCTION

The grand partition function plays a central role is statistical thermodynamics because all the properties of system with a variable number of particles can be expressed in terms of it. In order to express our results in terms of Z_G we shall first derive the entropy S, which is defined as

$$S = k\sum_{N,r} p_{N,r} \ln p_{N,r}$$

$$= -k\sum_{N,r} p_{N,r}\left[\beta \mu N_r - \beta E_{N,r} - \ln Z_G\right]$$

$$= -k\left[\mu \beta\sum_{N,r} p_{N,r}N_r - \beta\sum_{N,r} p_{N,r}E_{N,r} - \ln Z_G\right]$$

$$= -k\left[\mu\beta\overline{N} - \beta\overline{E} - \ln Z_G\right]$$

$$= \frac{\overline{E}}{T} - \frac{\mu\overline{N}}{T} + k\ln Z_G \tag{7.13.1}$$

where $\overline{E}$ *and* $\overline{N}$ are the mean energy and mean number of particles of the system. In a macroscopic system the fluctuations are usually negligibly small, so the mean values are just the actual values of these quantities. Accordingly we frequently omit bars from such quantities.

Other thermodynamic functions are expressed in terms of grand partition function are as follows.

Total energy $$E \text{ or } U = -\left(\frac{\partial \ln Z_G}{\partial \beta}\right)_{V,\mu} \tag{7.13.2}$$

Mean number of particles $$\overline{N} = \eta\left(\frac{\partial \ln Z_G}{\partial \eta}\right), \quad \text{where} \quad \eta = e^{\beta\mu} \tag{7.13.4}$$

Mean pressure of the system $$\Pi = \frac{1}{\beta}\left(\frac{\partial \ln Z_G}{\partial V}\right)_{\beta,\mu} \tag{7.13.5}$$

7.14 GRAND POTENTIAL Φ

We define *grand potential* Φ as

$$\Phi = \Phi(T,V,\mu) = -kT \ln Z_G(T,V,\mu) = E - TS - \mu N \tag{7.14.1}$$

[This quantity is similar to the Helmholtz function F, which is related to canonical partition function Z (*T*, *V*, *N*) as *F* = –kT ln *Z* (*T*, *V*, *N*) = *E* – *TS*.]

The other thermodynamic functions are related to grand potential as given below.

$$S = -\left(\frac{\partial \Phi}{\partial T}\right)_{V,\mu}, \quad \overline{N} = -\left(\frac{\partial \Phi}{\partial \mu}\right)_{V,T} \tag{7.14.2}$$

7.15 IDEAL GAS FROM GRAND PARTITION FUNCTION

The canonical partition function *Z*(*T*, *V*, *N*) for a gas of *N* particles contained in a volume *V* and at temperature *T* is given by

$$Z(T,V,N)=\frac{1}{N\,!}[z(T,V)]^N=\frac{V^N}{N\,!}\left(\frac{2\pi\ mkT}{h^2}\right)^{3N/2} \tag{7.15.1}$$

The grand partition function for the gas is

$$Z_G=Z_G(T,V,\mu)=\sum_{N=0}^{\infty}e^{\beta\mu N}Z(T,V,N)=\sum_{N=0}^{\infty}\frac{\left[z\ e^{\beta\mu}\right]^N}{N\,!} \tag{7.15.2}$$

Eq. (7.15.2) may be simplified using the property of exponential function.

$$e^x=1+\frac{x}{1!}+\frac{x^2}{2!}+.......=\sum_{n=0}^{\infty}\frac{x^n}{n!} \tag{7.15.3}$$

In view of Eq.(7.15.3) we can write (7.15.2) as

$$Z_G=\exp\left[z\,e^{\beta\mu}\right] \tag{7.15.4}$$

Let $\eta=e^{\beta\mu}$. Therefore

$$Z_G=\exp[z\eta] \tag{7.15.5}$$

Now grand potential for the gas is

$$\begin{aligned}\Phi &= -kT\ln Z_G\\ &= -kT\,z\eta\\ &= -kT\left(\frac{2\pi\ mkT}{h^2}\right)^{3/2}V\ e^{\beta\mu}\end{aligned} \tag{7.15.6}$$

The entropy of the gas is

$$S=-\left(\frac{\partial\Phi}{\partial(kT)}\right)_{V,\mu}=V\left(\frac{(2\pi\ mkT)^{3/2}}{h^3}\right)e^{\beta\mu}\left(\frac{5}{2}-\beta\ \mu\right) \tag{7.15.7}$$

This is Sakur-Tetrode equation

The mean number of particles in the system is

$$\overline{N}=-\left(\frac{\partial\Phi}{\partial\mu}\right)_{V,T}=V\frac{(2\pi\,mkT)^{3/2}}{h^3}e^{\beta\mu}=-\frac{\Phi}{kT}$$

From this equation we can get chemical potential

$$\mu=-kT\ \ln\left[\frac{V}{\overline{N}}\frac{(2\pi\ mkT)^{3/2}}{h^3}\right]=-kT\ \ln\frac{z_G}{\overline{N}} \tag{7.15.8}$$

Thus the chemical potential increases with increase in concentration of particles.

The pressure of the gas is given by

$$\Pi = -\left(\frac{\partial \Phi}{\partial V}\right)_{T,\mu} = kT\left(\frac{\bar{N}}{V}\right) \tag{7.15.9}$$

This gives

$$\Pi V = \bar{N}kT$$

which is the perfect gas law.

7.16 OCCUPATION NUMBER OF AN ENERGY STATE FROM GRAND PARTITION FUNCTION : FERMI-DIRAC AND BOSE-EINSTEIN DISTRIBUTION

The state of a system is specified by a set of occupation number n_1, n_2, ...n_r, ...of single particle states with energies $\varepsilon_1 \leq \varepsilon_2 \leq ......\varepsilon_r \leq .....$ Consider a state of the system in which it contains N particles and total energy $E_{N,r}$, which are given by

$$N = \sum_r n_r \qquad and \qquad E_{N,r} = \sum_r n_r \varepsilon_r$$

Of course, in grand canonical ensemble, N is variable and free to take on values $N = 0, 1, 2,....$

The grand canonical partition function of the system is

$$Z_G = \sum_{N=0}^{\infty} \sum_{n_1, n_2, ...}^{(N)} e^{\beta \mu N} \, e^{-\beta E_{N,r}}$$

$$= \sum_{N=0}^{\infty} \sum_{n_1, n_2 ...}^{(N)} e^{\beta \mu (n_1 + n_2 + ...)} \, e^{-\beta (n_1 \varepsilon_1 + n_2 \varepsilon_2 + ...)} \tag{7.16.1}$$

The superscript N over the summation sign means that the occupation numbers obeys the condition $n_1 + n_2 + ... = N$. In the above summation first sum over all the values of n_1, n_2 ....for a fixed value of N and then sum over all the values of N from $N = 0$ to ∞. This way of summation is equivalent to summation over all values of n_1, n_2, ... independently of each other. So the above expression can be expressed as

$$Z_G = \sum_{n_1} \sum_{n_2} .... e^{\beta \mu (n_1 + n_2 + ...)} \, e^{-\beta (n_1 \varepsilon_1 + n_2 \varepsilon_2 + ...)}$$

$$= \left(\sum_{n_1 = 0}^{\infty} e^{\beta (\mu - \varepsilon_1) n_1} \right) \left(\sum_{n_2 = 0}^{\infty} e^{\beta (\mu - \varepsilon_2) n_2} \right) (......)(.........)$$

$$= \prod_r \left(\sum_{n_r}^{\infty} e^{\beta (\mu - \varepsilon_r) n_r} \right) \tag{7.16.2}$$

$$= \prod_r (Z_G)_r \tag{7.16.3}$$

where $(Z_G)_r$ is given by

$$(Z_G)_r = \sum_{n_r} e^{\beta (\mu - \varepsilon_r) n_r} \tag{7.16.4}$$

If the system consists of *fermions*, then n_r = 0 or 1 and therefore

$$(Z_G)_r = \sum_{n_r=0,\text{or }1} e^{\beta(\mu-\varepsilon_r)n_r} = \left(1+e^{\beta(\mu-\varepsilon)}\right) \tag{7.16.5}$$

Hence the grand partition function for *fermions is*

$$Z_G = \prod_r \left(1+e^{\beta(\mu-\varepsilon)}\right) \qquad (\textit{fermions}) \tag{7.16.6}$$

The *grand potential* for a system of *fermions* is

$$\begin{aligned}\Phi_{fermions} &= -kT\ln Z_G \\ &= -kT\sum_r \ln\left(1+e^{\beta(\mu-\varepsilon_r)}\right) \\ &= \sum_r -kT\ln\left(1+e^{\beta(\mu-\varepsilon_r)}\right) \qquad (7.16.7)\\ &= \sum_r \phi_r \qquad (7.16.8)\end{aligned}$$

The mean occupation number of r-th energy state is

$$\begin{aligned}\bar{n}_r &= -\frac{\partial\phi_r}{\partial\mu} \qquad (7.16.9)\\ &= -\frac{\partial}{\partial\mu}\left[-kT\ln\left(1+e^{\beta(\mu-\varepsilon_r)}\right)\right] \\ &= \frac{e^{\beta(\mu-\varepsilon_r)}}{1+e^{\beta(\mu-\varepsilon_r)}} \\ &= \frac{1}{e^{\beta(\varepsilon_r-\mu)}+1} \qquad (7.16.10)\end{aligned}$$

This expression for the mean occupation number for state of energy ε_r is called *Fermi-Dirac distribution* for particles with half-integral spin (*fermions*).

If the system consists of *bosons*, then n_r = 0, 1, 2, 3, ... and therefore the grand partition function for particles with integral spin (*bosons*) is

$$\begin{aligned}Z_G &= \prod_r \left(\sum_{n_r=0,1,2,3,\ldots\ldots} e^{\beta(\mu-\varepsilon_r)n_r}\right) \\ &= \prod_r \left(1+e^{\beta(\mu-\varepsilon_r)}+e^{2\beta(\mu-\varepsilon_r}+\ldots\ldots\ldots\ldots\right) \\ &= \left(\frac{1}{1-e^{\beta(\mu-\varepsilon_r)}}\right) \qquad \mu < \varepsilon_r \qquad (7.16.11)\end{aligned}$$

The *grand potential* $\Phi = E - TS - \mu N$ for a system of bosons is

$$\Phi_{bosons} = -kT \ln Z_G$$

$$= -kT \sum_i \ln\left(1 - e^{\beta(\mu-\varepsilon_r)}\right)^{-1}$$

$$= kT \sum_r \ln\left(1 - e^{\beta(\mu-\varepsilon_r)}\right)$$

$$= \sum_r kT \ln\left(1 - e^{\beta(\mu-\varepsilon_r)}\right) \tag{7.16.12}$$

$$= \sum_r \phi_r \tag{7.16.13}$$

The mean occupation number of state of energy ε_r is

$$\bar{n}_r = -\frac{\partial \phi_r}{\partial \mu} = -\frac{\partial}{\partial \mu}\left[kT \ln\left(1 - e^{\beta(\mu-\varepsilon_r)}\right)\right]$$

$$= \frac{e^{\beta(\mu-\varepsilon_r)}}{1 - e^{\beta(\mu-\varepsilon_r)}}$$

or

$$\bar{n}_r = \frac{1}{e^{\beta(\varepsilon_r-\mu)} - 1} \tag{7.16.14}$$

This is the Bose-Einstein distribution function for a system of particles with integral spin (*bosons*).

For a system of bosons (such as photons, phonons, etc.) whose number is not conserved, the chemical potential μ is zero. The Bose-Einstein distribution for such particles is

$$\bar{n}_r = \frac{1}{e^{\beta\varepsilon_r} - 1}, \quad \varepsilon_r = \hbar\omega_r \quad \text{(for photons and phonons)} \tag{7.16.15}$$

8

Application of Partition Function

8.1 SPECIFIC HEAT OF SOLIDS

8.1.1 Einstein Model

In 1907 Einstein made an attempt to explain the temperature-dependence of the heat capacity of solids on the basis of quantum theory, and made the following assumptions.

1. The atoms in a solid vibrate about their fixed lattice sites. Their vibrations are independent of each other and the frequency ω of vibration, called Einstein frequency, is the same for all atoms, and is a characteristic constant of the solid. The vibration of an atom can be resolved into three mutually independent equations say along the three Cartesian coordinate axes x, y and z. Thus a solid containing N_0 atoms is equivalent to 3 N_0 harmonic oscillators.
2. An oscillator of frequency ω can have only distance values of energy* given by discrete

$$\varepsilon = \left(n + \frac{1}{2}\right)\hbar\,\omega \quad \text{where } n = 0,\ 1,\ 2,\ 3,\ \ldots.. \tag{8.1.1}$$

The energy $\frac{1}{2}\hbar\,\omega$ is called the zero-point energy.

The partition function of one oscillator is

$$z = \sum_{r=0}^{\infty} e^{-\beta\,\varepsilon_r} = \sum_{r=0}^{\infty} e^{-\beta\hbar\omega\left(r+\frac{1}{2}\right)} = e^{-\frac{1}{2}\beta\hbar\omega} \sum_{r=0}^{\infty} e^{-\beta\hbar\omega r} \tag{8.1.2}$$

Let $x = \beta\hbar\omega = \dfrac{\hbar\omega}{kT}$. The expression for partition function becomes

$$z = e^{-x/2} \sum_{r=0}^{\infty} e^{-x\,r}$$

$$= e^{-x/2}\left[1 + e^{-x} + e^{-2x} + \ldots\ldots\right]$$

$$= e^{-x/2}\left[\frac{1}{1-e^{-x}}\right] \tag{8.1.3}$$

The Helmholtz free energy for one oscillator is

$$\begin{aligned} F_1 &= -kT \ln z \\ &= -kT\left[-\frac{x}{2} - \ln\left(1-e^{-x}\right)\right] \\ &= \left[\frac{1}{2}\hbar\omega + \frac{1}{\beta}\ln\left(1-e^{-\beta\hbar\omega}\right)\right] \end{aligned} \tag{8.1.4}$$

The average energy of an oscillator is

$$\begin{aligned} \bar{\varepsilon} &= -\frac{\partial \ln z}{\partial \beta} = -\frac{\partial}{\partial \beta}\left(-\frac{F_1}{kT}\right) = \frac{\partial}{\partial \beta}(\beta F_1) \\ &= \frac{\partial}{\partial \beta}\left[\frac{1}{2}\beta\hbar\omega + \ln\left(1-e^{-\beta\hbar\omega}\right)\right] \\ &= \frac{1}{2}\hbar\omega + \frac{\hbar\omega}{e^{\beta\hbar\omega}-1} \end{aligned} \tag{8.1.5}$$

The energy of crystal containing N atoms is

$$E = 3N\bar{\varepsilon} = 3N\left[\frac{1}{2}\hbar\omega + \frac{\hbar\omega}{e^{\beta\hbar\omega}-1}\right] \tag{8.1.6}$$

The specific heat at constant volume is

$$\begin{aligned} C_V &= \left(\frac{\partial E}{\partial T}\right)_V = \left(\frac{\partial E}{\partial \beta}\right)\left(\frac{\partial \beta}{\partial T}\right) \\ &= \frac{3N}{kT^2}(\hbar\omega)^2 \frac{e^{\beta\hbar\omega}}{\left(e^{\beta\hbar\omega}-1\right)^2} \\ &= 3Nk\left(\frac{\hbar\omega}{kT}\right)^2 \frac{e^{\beta\hbar\omega}}{\left(e^{\beta\hbar\omega}-1\right)^2} \end{aligned} \tag{8.1.7}$$

Let us express this result in terms of Einstein temperature Θ_E defined by $\hbar\omega = k\Theta_E$.

$$C_V = 3Nk\left(\frac{\Theta_E}{T}\right) \frac{e^{\Theta_E/T}}{\left(e^{\Theta_E/T}-1\right)^2} \tag{8.1.8}$$

(*i*) At high temperature $T >> \Theta_E$,

$$e^{\Theta_E/T} = 1 + \frac{\Theta_E}{T} + \ldots\ldots\ldots\ldots$$

The expression for C_V simplifies to

$$C_V = 3Nk\left(\frac{\Theta_E}{T}\right)^2 \frac{1+\Theta_E/T}{(\Theta_E/T)^2} = 3Nk\left(1+\frac{\Theta_E}{T}\right) \tag{8.1.9}$$

As $T \to \infty$, $C_V = 3Nk = 3R$

Thus at high temperature the heat capacity of solid is independent of temperature and is equal to $3R$ which is the Dulong-Petits law.

(*ii*) At low temperature Θ_E/T is large and so is $\exp(\Theta_E/T)$. The expression for heat capacity may be simplified as follows.

$$C_V = 3Nk\left(\frac{\Theta_E}{T}\right)^2 \frac{e^{\Theta_E/T}}{\left(e^{\Theta_E/T}\right)^2} = 3Nk\left(\frac{\Theta_E}{T}\right)^2 \frac{1}{e^{\Theta_E/T}}$$

$$= 3Nk \frac{1}{\left(\frac{T}{\Theta_E}\right)^2\left[1+\frac{\Theta_E}{T}+\frac{1}{2!}\left(\frac{\Theta_E}{T}\right)^2+\frac{1}{3!}\left(\frac{\Theta_E}{T}\right)^3+\ldots\ldots\right]}$$

$$= 3Nk\frac{1}{\left(\frac{T}{\Theta_E}\right)^2+\frac{T}{\Theta_E}+\frac{1}{2!}+\frac{1}{3!}\frac{\Theta_E}{T}+\ldots.} \tag{8.1.10}$$

As $T\to 0$, $C_V \to 0$. The heat capacities computed on the basis of Einstein's model at different temperatures are, in general, in good agreement with the experimental results. However, careful and detailed comparison of the predictions from Einstein's theory with experimental results shows that the agreement is only approximate. It has also been shown by the experiments that at low temperature, the heat capacity varies as T^3 whereas Einstein's theory predicts exponential variation of specific heat with temperature. This difference between theory and the experiment is not very surprising because the model assumed by Einstein is over simplified.

For example, Einstein has assumed that the atoms in the solid are vibrating independent of each other, which can not be realistic, as atoms are strongly coupled with each other and therefore if an atom is disturbed it must effect the other atoms. Consequently one must consider the vibrations of the groups of atom rather than the vibration of each atom independently. Moreover, Einstein has also assumed that all the atoms vibrate with a single frequency ω,, which is again not correct. In order to develop a more appropriate theory, which can provide a realistic explanation of the observed experimental facts, Debye assumed a more realistic model of solid and provided a better theory of specific heat of solid.

8.1.2 Debye Model

The atoms in a solid cannot be assumed to be independent. We must take into account their cooperative interactions. Taking this interaction into consideration, we arrive at a theory of heat capacity that is in agreement with experimental results. At 0 K, the system of atoms comprising a solid is in the ground state having the minimum energy. After receiving energy from outside, an atom moves in a certain direction from its equilibrium position. A force striving to return it to the equilibrium position develops. Therefore while leaving the equilibrium position, the atom exerts a certain force on neighbouring atoms, which in turn are to leave their equilibrium positions, as a result of which the motion becomes cooperative. This cooperative motion, when the displacement of one atom is transferred to the neighbouring atom and then to the next neighbour and so on, is an acoustic wave in the solid.

Taking into account the interaction between atoms, a system of atoms must be considered as a set of coupled oscillators. In this case, any motion of the system of atoms can be represented as a superposition of normal oscillations or normal modes of the system. Each normal mode is characterized by its frequency ω and the energy ε of this mode is given by $\varepsilon = \hbar\omega$.

A solid can support longitudinal and transverse waves both with different velocities. Transverse mode may have two different directions of polarization. There is a standard method of calculating the number of modes for each polarization in an isotropic solid. The number of modes of acoustic oscillations in the frequency range $d\omega$ at frequency ω, in a solid of volume V is given by

$$g(\omega)d\omega = \frac{V}{2\pi^2 v^3}\omega^2 d\omega \tag{8.1.11}$$

where v is velocity of the wave. In a isotropic solid the velocities of the two transverse waves are equal. Taking the longitudinal and transverse mode both, the number of modes in the frequency interval $d\omega$ about ω is given by

$$g(\omega)d\omega = \frac{V}{2\pi^2}\left(\frac{1}{v_l^3} + \frac{2}{v_t^3}\right)\omega^2 d\omega \tag{8.1.12}$$

The factor 2 multiplying in $1/v_t^2$ is due to the fact that there are two independent transverse directions of polarization. We define a mean velocity v by

$$\frac{3}{v^3} = \frac{1}{v_l^3} + \frac{2}{v_t^3} \tag{8.1.13}$$

According to Debye, a crystal with N atoms possesses $3N$ modes in all. The maximum frequency ω_D is such that there are $3N$ modes altogether *i.e.*,

$$\int_0^{\omega_D} g(\omega)d\omega = 3N \tag{8.1.14}$$

The cut-off frequency ω_D occurs because at sufficiently high frequency *i.e.*, short wavelength, we cannot ignore the atomic nature of the solid. That is at short wavelengths, the solid is no longer acts as a continuum. A crystal with inter-atomic spacing a cannot propagate waves with wavelengths less than $\lambda_{min} = 2a$. In this case neighbouring atoms vibrate in anti-phase.

Let us introduce a characteristic temperature Θ_D related to Debye cut-off frequency ω_D defined by

$$\hbar\omega_D = k\Theta_D \quad or \quad \Theta_D = \frac{\hbar\omega}{k} \tag{8.1.15}$$

So Debye condition (8.1.14) in terms of energy $\varepsilon = \hbar\omega$ becomes

$$\int_0^{k\Theta_D} g(\varepsilon)d\varepsilon = 3N$$

$$\frac{3V}{2\pi^2 v^3 \hbar^3}\int_0^{k\Theta_D} \varepsilon^2 d\varepsilon = 3N$$

$$\frac{3V}{2\pi^2 v^3 \hbar^3} = \frac{9N}{(k\Theta_D)^3} \tag{8.1.16}$$

With the help of Eq. (8.1.12), (8.1.13) and (8.1.14) we can write

$$g(\varepsilon)d\varepsilon = \frac{9N}{(k\Theta_D)^3}\varepsilon^2 d\varepsilon \tag{8.1.17}$$

The average number of normal modes (oscillators) with energy ε, are given by

$$\bar{n}(\varepsilon) = \frac{1}{e^{\varepsilon/kT}-1} \tag{8.1.18}$$

The energy of the crystal is

$$E = \int_0^{k\Theta_D} \varepsilon\, g(\varepsilon)\bar{n}(\varepsilon)\, d\varepsilon$$

$$= \frac{9N}{(k\Theta_D)^3}\int_0^{k\Theta_D} \frac{\varepsilon^3 d\varepsilon}{e^{\varepsilon/kT}-1}$$

$$= 9NkT\left(\frac{T}{\Theta_D}\right)^3 \int_0^{\Theta_D/T} \frac{x^3 dx}{e^x - 1}, \tag{8.1.19}$$

where $x = \frac{\hbar\omega}{kT} = \frac{\varepsilon}{kT}$, $x_D = \frac{\hbar\omega_D}{kT} = \frac{\Theta_D}{T}$

(*i*) At high temperature ($T >> \Theta_D$),

$$\frac{x^3}{e^x - 1} = \frac{x^3}{\left(1 + \frac{x}{1} + ...\right) - 1} \approx x^2 \quad \text{as } x \text{ is small.}$$

Therefore (8.1.19) simplifies to

$$E = \frac{9NkT^4}{\Theta_d^3}\left(\frac{x^3}{3}\right)_0^{\Theta_D/T} = 3NkT \tag{8.1.20}$$

The molar heat capacity at constant volume is

$$C_V = \left(\frac{\partial E}{\partial T}\right)_v = 3Nk = 3R \tag{8.1.21}$$

(*ii*) At low temperature, T << Θ_D, $x \to \infty$ and the integral

$$\int_0^{x_D} \frac{x^3}{e^x - 1} dx = \int_0^{\infty} \frac{x^3}{e^x - 1} dx = \frac{\pi^4}{15}$$

The energy of the system is

$$E = \frac{9NkT^4}{\Theta_D^3}\left(\frac{\pi^4}{15}\right) \tag{8.1.22}$$

The heat capacity

$$C_V = \left(\frac{\partial E}{\partial T}\right)_V = \frac{9Nk}{\Theta_D^3}\left(\frac{\pi^4}{15}\right)T^3 \tag{8.1.23}$$

Thus the heat capacity of solid at low temperature varies as T^3 . This law is known as Debye T^3 law.

8.2 PHONON CONCEPT

The energy $\varepsilon = \hbar\omega$ corresponding to a mode of acoustic vibration with frequency ω in a solid suggests that a mode should be treated as a *quasiparticle*. Such a *quasiparticle* associated with the modes of acoustic oscillations is called a *phonon*. The introduction of the phonon concept is a fruitful approach, which considerably simplifies the reasoning. The thermal vibrations of a lattice are equivalent to an aggregate of phonons and the latter may be treated as an ideal bose gas. The phonon is a *quanta* of acoustic wave with energy $\varepsilon = \hbar\omega$ and momentum $p = \frac{\varepsilon}{v}$, where v is mean velocity of acoustic wave in solid and is given by $\frac{3}{v^3} = \frac{1}{v_l^3} + \frac{2}{v_t^3}$, v_l and v_t are the velocities of the longitudinal and transverse waves.

The density of states for a gas of phonons is given by

$$g(p)dp = \frac{3V}{h^3} 4\pi\, p^2 dp \tag{8.2.1}$$

The factor 3 takes into account of three possible polarizations of phonon. V is volume of the solid. Making use of the relation $\varepsilon = v\,p$, the expression for the density of states can be transformed in terms of energy.

$$g(\varepsilon)d\varepsilon = \frac{3V}{2\pi^2 v^3 \hbar^3}\varepsilon^2 d\varepsilon \tag{8.2.2}$$

Let the maximum frequency of the acoustic oscillations be ω_D and the energy of the corresponding phonon be $\hbar\omega_D$. We define a characteristic temperature Θ_D such that $\hbar\omega_D = k\Theta_D$ *i.e.*, $\Theta_D = \frac{\hbar\omega}{k}$. The total number of modes of oscillation in the solid containing N atoms is $3N$. Thus

$$\int_0^{k\Theta_D} g(\varepsilon)d\varepsilon = 3N$$

$$\frac{3V}{2\pi^2 v^3 \hbar^3}\int_0^{k\Theta_D} \varepsilon^2 d\varepsilon = 3N$$

$$g(\varepsilon) = \frac{9N}{(k\Theta_D)^3}e^2 \tag{8.2.3}$$

The phonons are bosons and their number is not conserved so the occupation number of phonons in the state with energy ε is given by

$$\bar{n}(\varepsilon) = \frac{1}{e^{\varepsilon/kT} - 1} \tag{8.2.4}$$

The energy of phonons in the solid is

$$E = \int_0^{k\Theta_D} \varepsilon.g(\varepsilon).\bar{n}(\varepsilon).d\varepsilon$$

$$= \frac{9N}{(k\Theta_D)^3}\int_0^{k\Theta_D} \frac{\varepsilon^3}{e^{\varepsilon/kT} - 1}d\varepsilon$$

$$= 9NkT\left(\frac{T}{\Theta_D}\right)^3 \int_0^{\Theta_D/T} \frac{x^3}{e^x - 1}dx, \tag{8.2.5}$$

where $$x = \frac{\hbar\omega}{kT} = \frac{\varepsilon}{kT}.$$

At high temperature x is small and $\frac{x^3}{e^x - 1} \cong x^2$.. So Eq. (8.2.5) simplifies to

$$E = 3NkT$$

and hence

$$C_V = 3Nk = 3R \quad \text{(Dulong-Petits Law)}$$

At low temperature, x is very large, and the integral in Eq. (8.2.5)) is given by

$$\int_0^{\Theta_D/T} \frac{x^3}{e^x - 1} dx = \int_0^{\infty} \frac{x^3}{e^x - 1} dx = \frac{\pi^4}{15}$$

Therefore

$$E = \frac{9NkT^4}{\Theta_D^3}\left(\frac{\pi^4}{15}\right) \tag{8.2.6}$$

Heat capacity

$$C_V = \left(\frac{3Nk\pi^4}{5\Theta_D{}^3}\right) T^3 \tag{8.2.7}$$

Thus the heat capacity of a solid varies as T^3 at low temperature. This is Debye T^3 law.

8.3 PLANCK'S RADIATION LAW: PARTITION FUNCTION METHOD

Let us denote the single particle states of photon gas by 1, 2, 3, ...r... with energies ε_1, ε_2, ε_3,...... e_r .....and occupation numbers n_1, n_2, n_3, ...n_r ....For photons n_r = 0, 1, 2, ...for all r. *i.e.*, each of the occupation number assumes all possible values 0, 1, 2, ...independent of the values of the other occupation numbers. The partition function of the photon gas is

$$Z = \sum_{n_1, n_2, \ldots} e^{-\beta \sum_{r=1}^{\infty} n_r \varepsilon_r} \tag{8.3.1}$$

where $\sum\limits_{n_1, n_2 \ldots}$ stands for summation over all sets of occupation numbers. Eq, (8.3.1)) can be written as

$$Z = \sum_{n_1}^{\infty}\sum_{n_2}^{\infty} \ldots\ldots\ldots e^{-\beta(n_1\varepsilon_1 + n_2\varepsilon_2 + \ldots.)} \tag{8.3.2}$$

$$= \left(\sum_{n_1}^{\infty} e^{-\beta n_1 \varepsilon_1}\right)\left(\sum_{n_2}^{\infty} e^{-\beta n_2 \varepsilon_2}\right)(\ldots\ldots.)(\ldots\ldots.)(\ldots\ldots.)$$

$$= \left(\frac{1}{1 - e^{-\beta\varepsilon_1}}\right)\left(\frac{1}{1 - e^{-\beta\varepsilon_2}}\right)(\ldots\ldots)(\ldots\ldots.)$$

$$= \prod_{r=1}^{\infty}\left(\frac{1}{1 - e^{-\beta\varepsilon_r}}\right) \tag{8.3.3}$$

Therefore

$$\ln Z = -\sum_{r=1}^{\infty} \ln\left(1 - e^{-\beta\varepsilon_r}\right) \tag{8.3.4}$$

where $\varepsilon_r = \hbar\omega$

The mean occupation number of photons in the state r is

$$\bar{n}_r(\omega) = -\frac{1}{\beta}\frac{\partial \ln Z}{\partial \varepsilon_r} = \frac{1}{e^{\beta \varepsilon_r} - 1} = \frac{1}{e^{\beta \hbar \omega} - 1} \tag{8.3.5}$$

The degeneracy of energy levels ε_r is given by

$$g(\omega)d\omega = 2\left(\frac{V}{2\pi^2 c^3}\omega^2\right)d\omega \tag{8.3.6}$$

The factor 2 accounts for the two independent directions of polarization of photon. Therefore

$$\ln Z = -\sum_r \left(\frac{V}{\pi^2 c^3}\omega^2\right)\ln\left(1 - e^{-\beta\hbar\omega_r}\right)$$

$$= -\frac{V}{\pi^2 c^3}\int_0^\infty \omega^2 \ln(1 - e^{-\beta\hbar\omega} d\omega \tag{8.3.7}$$

$$= -\frac{V}{\pi^2 c^3}\left[\frac{\omega^3}{3}\ln(1 - e^{-\beta\hbar\omega})\right]_0^\infty + \frac{V\beta\hbar}{3\pi^2 c^3}\int_0^\infty \frac{\omega^3 e^{-\beta\hbar\omega}}{1 - e^{-\beta\hbar\omega}} d\omega$$

$$= 0 + \frac{V\beta\hbar}{3\pi^2 c^3}\int_0^\infty \frac{\omega^3}{e^{\beta\hbar\omega} - 1} d\omega$$

$$= \frac{V}{3\pi^2 c^3(\beta\hbar)^3}\int_0^\infty \frac{x^3}{e^x - 1} dx \qquad \text{where} \qquad \beta\hbar\omega = x$$

$$= \frac{V\pi^2}{45}\left(\frac{kT}{\hbar c}\right)^3 \tag{8.3.8}$$

Average energy $$\bar{E} = -\frac{\partial \ln Z}{\partial \beta} = kT^2 \frac{\partial \ln Z}{\partial T} = \left(\frac{V\pi^2 k^4}{15\hbar^3 c^3}\right)T^4 \tag{8.3.9}$$

Mean pressure $$\bar{P} = \frac{1}{\beta}\frac{\partial \ln Z}{\partial V} = \frac{\pi^2 k^4 T^4}{45\hbar^3 c^3} \tag{8.3.10}$$

$$\bar{P}V = \frac{1}{3}\bar{E} \tag{8.3.11}$$

Entropy $$S = k[\ln Z + \beta \bar{E}]$$

$$= \frac{4V\pi^2 k^4 T^3}{45\hbar^3 c^3} \tag{8.3.12}$$

Radiation density

$$u(\omega, T)d\omega = g(\omega).n(\omega)d\omega(\hbar\omega)$$

$$= \left(\frac{\omega^2 d\omega}{\pi^2 c^3}\right)\left(\frac{1}{e^{\beta\hbar\omega} - 1}\right)(\hbar\omega)$$

$$= \frac{\hbar\omega^3}{\pi^2 c^3} \frac{1}{e^{\beta\hbar\omega} - 1} d\omega \qquad (8.3.13)$$

This is the Planck radiation law.

Appendix

A-1 Evaluation of integral $\int_{-\infty}^{\infty} e^{-x^2} dx$

Let
$$I = \int_{-\infty}^{\infty} e^{-x^2} dx \tag{1}$$

Also
$$I = \int_{-\infty}^{\infty} e^{-y^2} dy \tag{2}$$

Multiplying (1) and (2)

$$I^2 = \int_{-\infty}^{\infty} \int_{-\infty}^{\infty} e^{-\left(x^2+y^2\right)} dxdy \tag{3}$$

The range of integration is over the entire x-y plane. The integral (3) can be evaluated in terms of polar coordinates (r, θ).

$$x = r\cos\theta, \quad y = r\sin\theta, \quad x^2 + y^2 = r^2 \quad \text{and} \quad dxdy = rdrd\theta$$

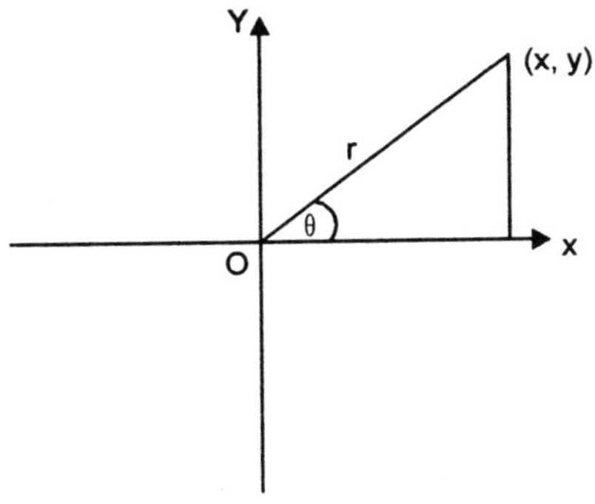

Fig. A-1 Relation between Cartesian and polar coordinates

The range of r is from 0 to ∞ and that of θ is from 0 to 2π. Therfore

$$I^2 = \int_0^\infty \int_0^{2\pi} r\, e^{-r^2}\, dr\, d\theta$$

$$= \int_0^\infty r\, e^{-r^2}\, dr \int_0^{2\pi} d\theta$$

$$= 2\pi \int_0^\infty r\, e^{-r^2}\, dr = \pi \int_0^\infty e^{-u}\, du = \pi \qquad (r^2 = u, \; 2\, r\, d\, r = du)$$

$$\therefore I = \sqrt{\pi}$$

Therefore $$\int_{-\infty}^{\infty} e^{-x^2}\, dx = \sqrt{\pi} \tag{4}$$

And $$\int_0^\infty e^{-x^2}\, dx = \frac{\sqrt{\pi}}{2} \tag{5}$$

A-2. Evaluation of Integral $I_n(a) = \int_0^\infty x^n e^{-ax^2}\, dx$

$$I_n(a) = \int_0^\infty x^n e^{-ax^2}\, dx \tag{1}$$

These integrals are found by calculating I_0 $(a)$and I_1 (a).

$$I_0(a) = \int_0^\infty e^{-ax^2}\, dx = \frac{1}{\sqrt{a}} \int_0^\infty e^{-y^2}\, dy, \qquad (\because ax^2 = y^2, \quad dx = \frac{1}{\sqrt{a}} dy)$$

$$I_0\,(a) = \frac{1}{2}\sqrt{\frac{\pi}{a}} \tag{2}$$

$$I_1(a) = \int_0^\infty x e^{-ax^2}\, dx = \frac{1}{2a} \int_0^\infty e^{-y}\, dy, \qquad where \;\; ax^2 = y, \;\; xdx = \frac{1}{2a} dy$$

$$= \frac{1}{2a}\left[-e^{-y}\right]_0^\infty = \frac{1}{2a}$$

$$I_1(a) = \frac{1}{2a} \tag{3}$$

$$I_n(a) = \int_0^\infty x^n e^{-ax^2}\,dx = -\frac{1}{2a}\int_0^\infty x^{n-1}\,d(e^{-ax^2})$$

$$= -\frac{1}{2a}\left[\left\{x^{n-1}e^{-ax^2}\right\}_0^\infty - (n-1)\int_0^\infty x^{n-2}e^{-ax^2}\,dx\right]$$

$$= 0 + \frac{n-1}{2a}\int_0^\infty x^{n-2}e^{-ax^2}\,dx$$

$$= \frac{n-1}{2a} I_{n-2}(a)$$

$$\therefore I_n(a) = \frac{n-1}{2a} I_{n-2}(a) \tag{4}$$

From (4)

$$I_2(a) = \frac{1}{4a}\sqrt{\frac{\pi}{a}}, \quad I_3(a) = \frac{1}{2a^2}, \quad I_4(a) = \frac{3}{8a^2}\sqrt{\frac{\pi}{a}}\ etc. \tag{5}$$

A-3. Evaluation of $I = \int_0^\infty \frac{x^3}{e^x - 1}\,dx$

This integral can be evaluated by expanding the integrand in a series. Since $e^x < 1$, throughout the range of integration, we can write

$$\frac{x^3}{e^x - 1} = \frac{e^{-x}x^3}{1 - e^{-x}} = e^{-x}.x^3\left[1 + e^{-x} + e^{-2x} + \ldots\ldots\right]$$

$$= \sum_{n=1}^{\infty} e^{-nx}\,x^3$$

Hence

$$I = \sum_{n=1}^{\infty} \int_0^{\infty} e^{-nx} . x^3 dx$$

$$= \sum_{n=1}^{\infty} \frac{1}{n^4} \int_0^{\infty} e^{-y} y^3 dy, \qquad \textit{where} \quad nx = y,$$

$$= \sum_{n=1}^{\infty} \frac{1}{n^4} \Gamma 4 = \sum_{n=1}^{\infty} \frac{1}{n^4} (3!) = 6 \sum_{n=1}^{\infty} \frac{1}{n^4} = 6\left(\frac{\pi^4}{90}\right) = \frac{\pi^4}{15}$$

A-4. Evaluation of $I = \int_0^{\infty} \frac{x^4 e^x}{(e^x - 1)^2} dx$

$$I = -\int_0^{\infty} x^4 \, d\left(\frac{1}{e^x - 1}\right) = -\left[\left(x^4 . \frac{1}{e^x - 1}\right)_0^{\infty} - 4 \int_0^{\infty} \frac{x^3}{e^x - 1} dx\right]$$

$$= 4 \int_0^{\infty} \frac{x^3}{e^x - 1} dx$$

$$= 4\left(\frac{\pi^4}{15}\right) = \frac{4}{15} \pi^4$$

A-5. Riemann Zeta Function ζ **(x)**

Riemann zeta function is defined as

$$\zeta(x) = 1 + \frac{1}{2^x} + \frac{1}{3^x} + \frac{1}{4^x} + \ldots.. \tag{1}$$

$$\zeta(2) = \sum_{n=1}^{\infty} \frac{1}{n^2} = \frac{\pi^2}{6} = 1.645$$

$$\zeta(4) = \sum_{n=1}^{\infty} \frac{1}{n^4} = \frac{\pi^4}{90} = 1.082$$

$$\sum_{n=1}^{\infty} \frac{(-1)^{n-1}}{n^2} = \frac{\pi^2}{12}$$

A-6. Stirling's Approximation

For large n, $\ln n! = n \ln n - n = n \ln (n/e)$ (1)

By definition $n! = 1.2.3.........n$

$$\ln n! = \ln 1 + \ln 2 + \ln 3 + ...\ln n$$

$$= \sum_{m=1}^{m} \ln m$$

This sum is exactly equal to the area under the step curve shown by broken line in the figure between $n = 1$ and $n = n$. This area may be approximated with the area under the smooth curve $y = \ln n$ between the same limits. For small values of n, the step curve differs appreciably from the smooth curve but the smooth curve becomes more and more nearly parallel to the n-axis as n increases.

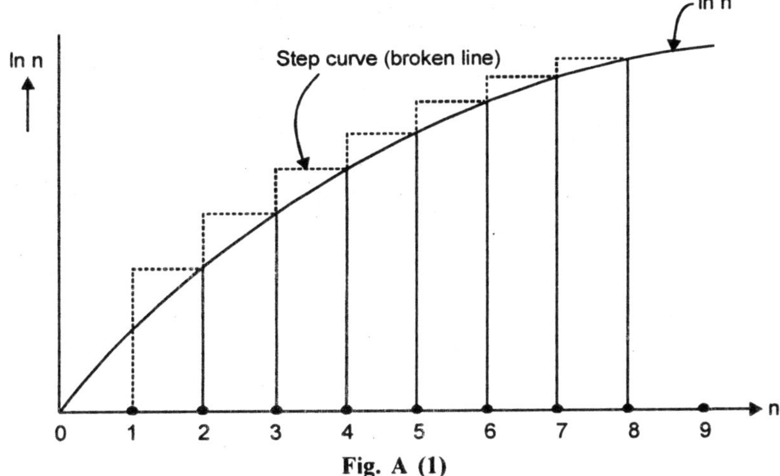

Fig. A (1)

$$\ln n! = \int_1^n \ln x dx = (x \ln x.)_1^n - \int_1^n x.\frac{1}{x}.dx$$

$$= n \ln n - n + 1$$

$$= n \ln n - n \tag{1}$$

(n >> 1, we can neglect 1)

An exact analysis gives the following series for n !.

$$n! = \sqrt{2\pi n}\left(\frac{n}{e}\right)^n \left[1 + \frac{1}{12n} + \frac{1}{288n^2} - \frac{139}{51840n^3} + ...\right]$$

Retaining the first term only we obtain

$$\ln n! = \frac{1}{2}\ln(2n\pi) + n \ln - n \tag{2}$$

n	n ln n !	ln − n	$\frac{1}{2}$ ln $2n\pi$
5	4.8	3.0	1.8
25	58.0	55.5	2.5
100	363.7	360.5	3.2

A-7. Gamma Function

Gamma function is defined as

$$\Gamma n = \int_0^\infty x^{n-1} e^{-x}\,dx, \quad n>0. \tag{1}$$

Now
$$\Gamma(n) = \int_0^\infty x^{n-1} e^{-x}\,dx = -\int_0^\infty x^{n-1}\,d(e^{-x})$$

$$= \left[-\left\{x^{n-1} e^{-x}\right\}_0^\infty - \int_0^\infty (n-1)x^{n-2}(-e^{x})\,dx\right]$$

$$= (n-1)\int_0^\infty x^{n-2} e^{-x}\,dx$$

$$= (n-1)\Gamma(n-1)$$

So
$$\Gamma(n) = (n-1)\Gamma(n-1) \tag{2}$$

For $n = ½$ we have

$$\Gamma\left(\frac{1}{2}\right) = \int_0^\infty x^{-1/2} e^{-x}\,dx = 2\int_0^\infty e^{-u^2}\,du = 2\frac{\sqrt{\pi}}{2} = \sqrt{\pi} \tag{3}$$

$$\Gamma(1) = \int_0^\infty e^{-x}\,dx = -\left(e^{-x}\right)_0^\infty = 1 \tag{4}$$

If n is an positive integer then

$$\begin{aligned}\Gamma(n) &= (n-1)\Gamma(n-1) = (n-1)(n-2)\Gamma(n-2)\\ &= (n-1)(n-2)(n-3)\ \ldots\ldots\ldots\ldots\ldots\Gamma(1)\\ &= (n-1)(n-2)(n-3)\ \ldots\ldots\ldots\ldots\ldots 1\\ &= (n-1)\,! \end{aligned} \tag{5}$$